Lecture Notes in Computer Science 16379

The series Lecture Notes in Computer Science (LNCS), including its subseries Lecture Notes in Artificial Intelligence (LNAI) and Lecture Notes in Bioinformatics (LNBI), has established itself as a medium for the publication of new developments in computer science and information technology research, teaching, and education.

LNCS enjoys close cooperation with the computer science R & D community, the series counts many renowned academics among its volume editors and paper authors, and collaborates with prestigious societies. Its mission is to serve this international community by providing an invaluable service, mainly focused on the publication of conference and workshop proceedings and postproceedings. LNCS commenced publication in 1973.

Altannar Chinchuluun · Rentsen Enkhbat ·
Ding-Zhu Du
Editors

Algorithmic Aspects in Information and Management

19th International Conference, AAIM 2025
Ulaanbaatar, Mongolia, June 23–25, 2025
Proceedings

Springer

Editors
Altannar Chinchuluun
National University of Mongolia
Ulaanbaatar, Mongolia

Rentsen Enkhbat
Mongolian Academy of Sciences
Ulaanbaatar, Mongolia

Ding-Zhu Du
University of Texas at Dallas
Richardson, TX, USA

ISSN 0302-9743 ISSN 1611-3349 (electronic)
Lecture Notes in Computer Science
ISBN 978-981-95-5656-4 ISBN 978-981-95-5657-1 (eBook)
https://doi.org/10.1007/978-981-95-5657-1

This Springer imprint is published by the registered company Springer Nature Singapore Pte Ltd.
The registered company address is: 152 Beach Road, #21-01/04 Gateway East, Singapore 189721, Singapore

Preface

This volume brings together a selection of papers presented at the 19th International Conference on Algorithmic Aspects in Information and Management (AAIM 2025), held in Ulaanbaatar, Mongolia, during June 23–25, 2025. The conference was organized by several higher education and research institutions including the Mongolian Academy of Sciences and National University of Mongolia. More than 60 speakers presented their works in person and online at the conference.

For these conference proceedings, we accepted 14 papers from 24 submissions after carefully evaluating their originality and quality. Each paper was reviewed in double-blind style and was evaluated by at least two independent members of the International Program Committee, consisting of many scholars from various countries and regions, distributed over the world, including Asia, North America, and South America. We would like to express our gratitude and appreciation to all the reviewers for their constructive comments on the papers.

We believe that this volume will be of great interest to graduate students and researchers in computer science, mathematics, and management science, as well as to engineers and scientists involved in algorithmic aspects of information and management science. We would like to thank all the authors and presenters who provided the research insights comprising the essential substance of the conference. We would also like to express our gratitude for the hard work of the local and international committee members for arranging the conference in a very professional and efficient manner.

October 2025

Altannar Chinchuluun
Rentsen Enkhbat
Ding-Zhu Du

Organization

General Chairs

Rentsen Enkhbat	Mongolian Academy of Sciences, Mongolia
Panos M. Pardalos	University of Florida, USA

Program Committee Chairs

Altannar Chinchuluun	National University of Mongolia, Mongolia
Ding-Zhu Du	University of Texas at Dallas, USA

Steering Chairs

Yinfeng Xu	Xi'an Jiaotong University, China
Binhai Zhu	Montana State University, USA

Program Committee

Ashwin Arulselvan	University of Strathclyde, UK
Milagros Baldemor	Don Mariano Marcos Memorial State University, Philippines
Hexi Baoyin	Tsinghua University, China
Saheya Barintag	Inner Mongolia Normal University, China
Sheng Bau	University of the Witwaterstand, South Africa
Radu Ioan Bot	University of Vienna, Austria
Alexander S. Buldaev	Buryat State University, Russia
Tiantian Chen	Santa Clara University, USA
Yongxi Cheng	Xi'an Jiaotong University, China
Battuvshin Chuluundorj	University of the Humanities, Mongolia
Donglei Du	University of New Brunswick, Canada
Hongwei Du	Harbin Institute of Technology, Shenzhen, China
Neng Fan	University of Arizona, USA
Ling Gai	University of Shanghai for Science and Technology, China
Smita Ghosh	Santa Clara University, USA

Munkhjargal Gochoo	United Arab Emirates University, UAE
Lu Han	Beijing University of Posts and Telecommunications, China
Masaru Kamada	Ibaraki University, Japan
Liying Kan	Shanghai University, China
Altangerel Lkhamsuren	German-Mongolian Institute for Resources and Technology, Mongolia
Bin Liu	Ocean University of China, China
Lkhagvadorj Munkhdalai	Empasoft Institute of Technology, Mongolia
Oyun-Erdene Namsrai	National University of Mongolia, Mongolia
Qiufan Ni	Guangdong University of Technology, China
Steffen Rebennack	Karlsruhe Institute of Technology, Germany
Alexander Strekalovsky	Institute for System Dynamics and Control Theory, Russia
Guangmo Tong	University of Delaware, USA
Ider Tseveendorj	Université de Versailles Saint-Quentin, France
Gantumur Tsogtgerel	National University of Mongolia, Mongolia
Olga Vasilieva	Universidad del Valle, Columbia
Richard Vogel	Farmingdale State College, USA
Chenchen Wu	Tianjin University of Technology, China
Petros Xanthopoulos	Stetson University, USA
Dachuan Xu	Beijing University of Technology, China
Ruidong Yan	IEIT SYSTEMS Co., Ltd
Wenguo Yang	University of Chinese Academy of Sciences, China
Huaming Zhang	University of Alabama in Huntsville, USA
Guochuan Zhang	Zhejiang University, China
Nan Zhang	Xidian University, China
Yapu Zhang	Beijing University of Technology, China
Zhao Zhang	Zhejiang Normal University, China
Binhai Zhu	Montana State University, USA
Jianming Zhu	University of Chinese Academy of Sciences, China

Local Committee

Ch. Ankhbayar	National University of Mongolia, Mongolia
B. Barsbold	National University of Mongolia, Mongolia
S. Batbileg	National University of Mongolia, Mongolia
B. Batgerel	Mongolian Academy of Sciences, Mongolia
N. Batnasan	National University of Mongolia, Mongolia

Ts. Batsukh	University of Finance and Economics, Mongolia
G. Battur	National University of Mongolia, Mongolia
D. Bayanjargal	National University of Mongolia, Mongolia
T. Bayartugs	Mongolian University of Science and Technology, Mongolia
J. Davaadulam	National University of Mongolia, Mongolia
J. Enkhbayar	National University of Mongolia, Mongolia
M. Mend-Amar	National University of Mongolia, Mongolia
S. Mengkezhula	National University of Mongolia, Mongolia
Ch. Naranchimeg	University of Finance and Economics, Mongolia
D. Nyamsuren	Mongolian University of Science and Technology, Mongolia
B. Otgonsuvd	Mongolian Academy of Sciences, Mongolia
B. Ser-Od	Mongolian University of Science and Technology, Mongolia
D. Tsedenbayar	Mongolian University of Science and Technology, Mongolia
D. Tseveennamjil	Mongolian University of Life Sciences, Mongolia
N. Tungalag	National University of Mongolia, Mongolia
V. Ulziibayar	Mongolian University of Science and Technology, Mongolia
D. Uuganbaatar	Mongolian Academy of Sciences, Mongolia

Contents

On the Twin Bridges Problem in Polygons

Haitao Jiang[1], Letu Qingge[2], Lusheng Wang[3], and Binhai Zhu[4](✉)

[1] School of Computer Science and Technology, Shandong University, Qingdao, China
htjiang@sdu.edu.cn

[2] Department of Computer Science,North Carolina Agricultural and Technical State University, Greensboro, NC 27411, USA
lqingge@ncat.edu

[3] Department of Computer Science, City University of Hong Kong, Hong Kong, SAR, China
cswangl@cityu.edu.hk

[4] Gianforte School of Computing, Montana State University, Bozeman, MT 59717, USA
bhz@montana.edu

Abstract. In this paper, we investigate the twin bridges problem in polygons, whose counterpart on trees has been studied recently. Given two disjoint simple polygons P and Q, each with at most n vertices, the problem is to find two vertices p_1 and p_2 in P (and q_1 and q_2 in Q) to build two bridges p_1q_1 and p_2q_2 such that the constrained diameter of the resulting geometric structure (e.g., a polygon with a hole when the two bridges do not intersect), i.e., the maximum of the shortest distance between two vertices in P, Q, or just in one of them, through at least one of the bridges, is minimized. The main results are summarized as follows: (1) If P and Q are arbitrary polygons, the problem can be solved in $O(n^6 \log n)$ time. (2) These results hold even if the bridges between p_i and q_i, $i \in \{1, 2\}$, are geodesic. (3) We show the general problem is NP-hard: given m disjoint polygons $P_1, ..., P_m$, add two bridges between P_i and P_{i+1}, $i \in [m-1]$, such that the constrained diameter of the resulting geometric structure (ideally, a polygon with multiple holes, but could be more complex if some of the bridges intersect) is minimized.

Keywords: Optimal bridge · Geometric algorithms · Max-Heaps · NP-completeness

1 Introduction

The optimal bridge problem between convex polygons was first studied by Cai, Xu and Zhu in 1999 [5]. The problem is to connect two convex polygons P and Q (with vertex set $V(P)$ and $V(Q)$ respectively), by adding a bridge pq, with $p \in V(P), q \in V(Q)$, such that the constrained diameter of the resulting polygon (if we double the bridge), which must go through the bridge, is minimized.

A. Chinchuluun et al. (Eds.): AAIM 2025, LNCS 16379, pp. 1–13, 2026.
https://doi.org/10.1007/978-981-95-5657-1_1

The problem attracts quite some attention which include improving the running time for this convex case [1,2,12,13], generalizing the polygons to non-convex [12,14,15], and even considering the case when the endpoints of the bridge, p and q, might not be visible to each other [3,4].

Motivated by the general problem of inserting edges to improve the stretch factor of a metric graph [8], recently Feng, Fernau and Zhu considered the problem of adding one or two edges between (geometric) trees where the objective is to minimize the constrained diameter [7]. The latter problem is called the *Twin Bridges* problem for which Feng, Fernau and Zhu gave an $O(n^4)$ time solution, where n is the maximum size of the two input trees.

In this paper, we consider the *Twin Bridges* problem in polygons P and Q, each with at most n vertices. The problem is to find two pairs of vertices (p_1, q_1) and (p_2, q_2) with $p_i \in V(P), q_i \in V(Q), i \in \{1, 2\}$, such that the constrained diameter of the resulting structure (if we double the two bridges and the bridges do not intersect, the structure is a polygon with one hole), which must go through one of the bridges, is minimized. We show that if P and Q each has at most n vertices, then the problem can be solved in $O(n^6 \log n)$ time. The solution uses an interesting technique involving partition two arrays with the same sets of indices, which could find some other applications. On the other hand, we show that the following generalized version is NP-hard: suppose we are given a sequence of m disjoint polygons $P_1, ..., P_m$, each with at most n vertices, and we need to connect twin bridges between P_i and P_{i+1}, for $i \in [m-1]$, so as to minimize the constrained diameter of the resulting geometric structure. We first go over several properties of the problem which would help us classify different variations of the problem. First of all, it was shown in [7] that the two bridges for trees can intersect. We show here that for polygons it is also the case; in fact, we give a possibly simplest example in Fig. 1. In Fig. 1 (I), we start with two degenerate polygons which are segments $P = p_1p_2, Q' = q'_1q'_2$ where $p_2 = q'_2$; moreover, $\triangle p_1p_2q'_1$ forms a regular triangle with edge length ℓ. Then we translate Q' vertically by a small vertical distance of ε, and now we have $Q = q_1q_2$. By triangle inequality, $|p_1q_2| = |p_2q_1| < |p_1p_2| + |p_2q_2| = \ell + \varepsilon$. In other words, if we choose bridges p_1q_2 and p_2q_1 they would incur an intersecting solution of value strictly less than $\ell + \epsilon$. On the other hand, the only non-intersecting solution is to have bridges p_1q_1 and p_2q_2, which incur a solution of an exact value $\ell + \varepsilon$.

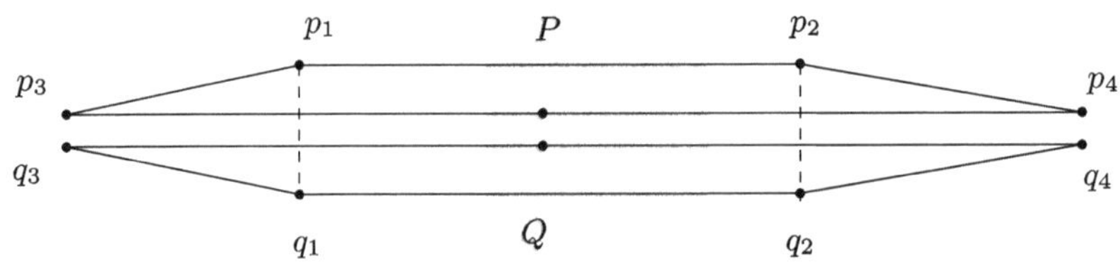

Fig. 1. In the twin bridges problem, the endpoints of two optimal bridges, e.g., p_1, q_1 and p_2, q_2 might not be visible to each other.

Secondly, in Fig. 1, we show that even when P, Q are convex, we could have a better solution p_1q_1 and p_2q_2 (compared with the solution composed of p_3q_3 and p_4q_4) such that p_i is not visible to q_i, for $i \in \{1, 2\}$. In reality, that means we could either build a high-rise straight bridge p_iq_i, or we could build geodesic bridges—in this case the geodesic bridge between p_i and q_i, $i \in \{1, 2\}$, is the shortest path between p_i and q_i treating P, Q as obstacles. We will focus on the former case first, which we call the Euclidean case; and at the end we briefly discuss the geodesic case.

The paper is organized as follows. In Sect. 2, we given some necessary definitions and solve a warm-up problem using max-heaps. In Sect. 3, we consider the Euclidean case of the twin bridges problem and then we briefly discuss the geodesic case of the twin bridges problem. In Sect. 4, we show the NP-hardness of the generalized problem. We conclude the paper in Sect. 5.

2 Preliminaries

2.1 Definitions

Given a natural number n, we use $[n]$ to represent the set $\{1, 2, ..., n\}$. Similarly, we use $[i, j]$ to represent the set $\{i, i+1, ..., j\}$ for $i \leq j$. Throughout the paper, we use P, Q to represent two non-intersecting simple polygons each of at most n vertices. (We will clearly state that they are convex, when it is the case.) We use $V(P)$ and $V(Q)$ (resp. $E(P)$ and $E(Q)$) to denote the vertices (resp. edges) in P and Q. A *bridge* between P and Q is a straight segment pq, $p \in V(P)$ and $q \in V(Q)$, whose Euclidean length is denoted as $|pq|$. (To reduce confusion, we avoid discussing geodesic bridges until at the end of Sect. 3.) Given two points s, t in P, the shortest path between s and t is denoted as $\delta_P(s, t)$ and when the context is clear we also use $\delta_P(s, t)$ to denote the length of the shortest path between s and t in P. It is well known that when P is triangulated, after a linear time preprocessing, $\delta_P(s, t)$ can be computed in $O(\log n)$ time [10].

The twin bridges problem is defined as follows: Given two polygons P and Q, compute two bridges p_1q_1 and p_2q_2, $p_i \in V(P)$ and $q_i \in V(Q)$, for $i \in \{1, 2\}$, such that the constrained diameter between $x \in V(P)$ and $z \in V(Q)$ through one of two the bridges, or between $x_1, x_2 \in V(P)$ (symmetrically, between $z_1, z_2 \in V(Q)$) through both of the bridges, is minimized. By a standard definition, the diameter between $V(P)$ and $V(Q)$ is the maximum of the minimum distance between $x \in V(P)$ and $z \in V(Q)$.

The readers might want to ask why in our definition, in addition to p_1, q_1, p_2 and q_2, x and z must also be vertices in P and Q respectively. (This question is natural as the input are polygons.) We try to answer this question by examining the following example in Fig. 2, where the two polygons P and Q are two line segments. In this case, we have no choice but picking vertices p_1, p_2, q_2 and q_2 as the endpoints for the bridges; in addition, we must list x and z as non-vertices, which means we need 2 variables to locate them. However, suppose the constrained diameter is determined by the only equation

$$|xp_1| + |p_1q_1| + |q_1z| = |xp_2| + |p_2q_2| + |q_2z|,$$

which involves two variables x and z. Hence we cannot obtain an exact solution for x and z. In fact, given x and z, it is always possible to arbitrarily slide x to x' and z to z' (in opposite directions) such that $|xx'| + |x'z'| = |xz| + |zz'|$ (x and z' would be the centers of an ellipse, and x' and z would be two points on that ellipse intersecting P and Q respectively).

In practice, this should not be a big problem as we could add a dense set of points on each edge, treating them as (pseudo-)vertices. That way, we could easily obtain a good approximation to the continuous problem (that p_1, q_1, p_2, q_2, x and z are arbitrary), although at a cost of increasing the running time.

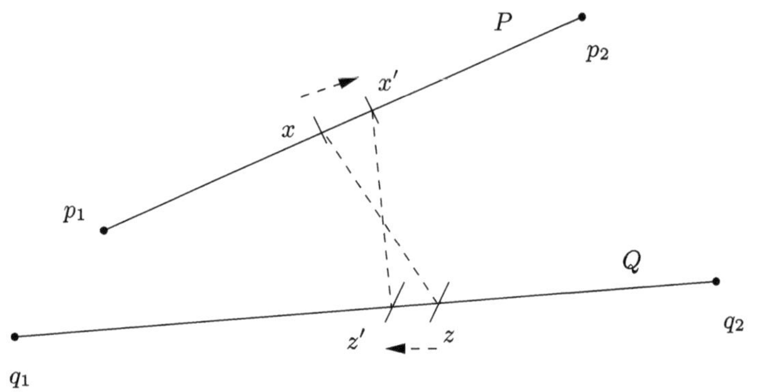

Fig. 2. Sliding x to x' and z to z', one can obtain another solution of the same value.

Coming back to our problem, we could classify the two cases in the definition as Case 1 and Case 2 respectively—depending on if the optimal solution is between vertices in $V(P)$ and $V(Q)$, or is between vertices only in $V(P)$ (or $V(Q)$). See Fig. 3 and Fig. 4 below. Note that x_i and z_i are the vertices of P and Q respectively.

2.2 A Warm-up Problem

Here we give an $O(n \log n)$ time solution for an interesting problem, which would be used as a subroutine for Case 1 of the twin bridges problem. Suppose we are given a sequence of n indices. There are two non-negative real functions f and g, where each $f(j)$ and $g(j)$, $j \in [1, n-1]$, takes constant time to compute; moreover, both of them can take a segment of indices $[1, i]$ or $[j, n]$ for $1 \leq i, j \leq n$, as parameters. In the latter case it takes $O(i)$ and $O(|n-j+1|)$ time to compute the maximum of $f(1), f(2), ..., f(i)$, which is $f(1, i) = \max\{f(1), f(2), ..., f(i)\}$, and the maximum of $f(j), f(j+1), ...f(n)$, which is $f(j, n) = \max\{f(j), f(j+1), ..., f(n)\}$. We can define $g(1, i)$ and $g(j, n)$ symmetrically. (*Hence, we could take f, g as arrays with n indices.*) The problem is to compute

$$F[i] = \min\{\max\{f(1, i), g(i+1, n)\}, \max\{f(i+1, n), g(1, i)\}\},$$

$i \in [n-1]$, which is a functional value obtained by breaking the index arrary $[1, n]$ at index i, and subsequently compute the corresponding min max-values. Eventually, we need to compute

$$F^*(n) = \min_{i \in [n-1]} F[i].$$

We define this as the *MinMax-Break* problem for two non-negative real functions f, g over an integer of n indices, or two non-negative real arrays $f[1, n]$ and $g[1, n]$. Certainly, the function f and g could vary among applications—given a sequence of n input points, in our application $f(i)$ is the distance between the i-th input point to a given point. With a brute-force method, we could just break each $i \in [n-1]$ and then compute each $F[i]$ in $O(n)$ time. It then takes a total of $\Theta(n^2)$ time to compute $F^*(n)$.

We show below, with two max-heaps each of at most $n-1$ values, that this problem can be solved in $O(n \log n)$ time. To achieve the goal, we define 4 heaps of size $n-1$, $H_1[i], H_2[i], G_1[i]$ and $G_2[i]$, with $i \in [n-1]$, as follows.

1. $H_1(i)$ is a max-heap on $\{f(1), f(2)..., f(i)\}$,
2. $H_2(i)$ is a max-heap on $\{f(i+1), f(i+2), ..., f(n)\}$,
3. $G_1(i)$ is a max-heap on $\{g(1), g(2)..., g(i)\}$,
4. $G_2(i)$ is a max-heap on $\{g(i+1), g(i+2), ..., g(n)\}$.

Note that these max-heap can be built if we take $f(-)$ and $g(-)$ simply as arrays; in fact, using the standard techniques as depicted in [6]. We only present the details on constructing and maintaining the heaps $H_1(i)$ and $H_2(i)$, extracting their maximum values (and the corresponding indices). It is symmetric for $G_1(i)$ and $G_2(i)$.

We start with $i = 1$, then $F_1(1)$ is a heap of size one, $\max(F_1(1)) \leftarrow f(1)$, and the index k_1 achieving the maximum value in the heap is $k_1 \leftarrow 1$. For $F_2(1)$, it is a heap over $f(2), f(3), ..., f(n)$, which can be constructed in $O(n)$ time [6]. Moreover, $\max(F_2(i))$ and the index k_2 achieving the maximum value can be computed in $O(\log n)$ time.

Now we show how to update the two heaps when $i \leftarrow i+1$. Given $F_1(i)$, we simply insert a new element $f(i+1)$ into it, then we run the standard `Heapify(i+1)` process to have $H_1(i+1)$, whose maximum value $\max(H_1(i+1))$ and the corresponding index k_1 can be computed in $O(\log n)$ time. Given $F_2(i)$, to compute $F_2(i+1)$, we need to delete $f(i+1)$ from $F_2(i)$ then try to maintain the heap property. Alternatively, we can change $f(i+1)$ to -1 and then run `Heapify(i+1)` to maintain the heap property. (The total size of $H_2(i+1)$ does not change, but the number of non-negative items decreases by one.) Nonetheless, we could return $\max(H_2(i+1))$ and its corresponding index k_2 in $O(\log n)$.

Assuming that the heaps $G_1(i)$ and $G_2(i)$ (and their corresponding maximum values) can all be updated in $O(\log n)$ time, we can then re-define $F(i)$ as

$$F[i] = \min\{\max\{\max(H_1(i)), \max(G_2(i))\}, \max\{\max(H_2(i)), \max(G_1(i))\}\},$$

which can be computed in $O(\log n)$ time. Clearly, all $F[i]$, for $i \in [n-1]$, can be computed in $O(n \log n)$ time. Hence, the final solution

$$F^*(n) = \min_{i \in [n-1]} F[i],$$

can be computed in an additional $O(n)$ time. Therefore we have the following theorem.

Theorem 1. *The MinMax-Break problem over $[n]$ can be solved in $O(n \log n)$ time.*

We show a simple example for $n = 4$. Let $f[1..4] = (5, 2, 6, 3)$, $g[1..4] = (7, 1, 8, 4)$. Then the maximum values of the four heaps, for each $i \in [3]$, are: $H_1[1, 3] = (5, 5, 6)$, $H_2[1, 3] = (6, 6, 3)$, $G_1[1, 3] = (7, 7, 8)$, and $G_2[1, 3] = (8, 8, 4)$. Furthermore,

$$\begin{aligned}
F[1] &= \min\{\max\{\max(H_1[1]), \max(G_2[1])\}, \max\{\max(H_2[1]), \max(G_1[1])\}\} \\
&= \min\{\max\{5, 8\}, \max\{6, 7\}\} = 7, \\
F[2] &= \min\{\max\{\max(H_1[2]), \max(G_2[2])\}, \max\{\max(H_2[2]), \max(G_1[2])\}\} \\
&= \min\{\max\{5, 8\}, \max\{6, 7\}\} = 7, \\
F[3] &= \min\{\max\{\max(H_1[3]), \max(G_2[3])\}, \max\{\max(H_2[3]), \max(G_1[3])\}\} \\
&= \min\{\max\{6, 4\}, \max\{3, 8\}\} = 6.
\end{aligned}$$

Consequently, $F[1..3] = (7, 7, 6)$ and $F^*(4) = 6$. This is achieved at $i = 3$.

3 Solutions for the Twin Bridges Problem

In this section, we present details for the twin bridges problem. We first consider the case when the bridges are straight (or Euclidean). We ignore the visibility constraint of the endpoints of a bridge; in fact, as can be seen a bit later, it is easy to add the visibility constraint to have a solution fulfill the visibility requirement. The geodesic case will be discussed briefly after the Euclidean case.

3.1 When the Bridges Are Euclidean

In this subsection, we first cover the case when two bridges are Euclidean. We first have the following lemma.

Lemma 1. *Given two disjoint simple polygons P, Q each with at most n vertices, when the two bridges p_1q_1 and p_2q_2 are fixed ($p_i \in V(P), q_i \in V(Q), i \in \{1, 2\}$), Case 1 of the 2-bridge problem can be solved in $O(n^2 \log n)$ time.*

Proof. We refer to Fig. 3. Suppose, for Case 1, both x and z are vertices, i.e., $x \in V(P)$ and $z \in V(Q)$, and the optimal solution value is $\delta_P(x, p_1) + |p_1q_1| + \delta_Q(q_1, z)$. Let the vertices in P ordered in clockwise order from p_1 to p_2 be: $p_1(= x_0), x_1, x_2, ..., x_k, p_2(= x_{k+1})$, where $k \in O(n)$. The constrained diameter is obtained by identifying an edge $e_i = (x_i, x_{i+1})$ $(i \in [k-1])$ such that either some vertex x in the sequence $(x_1, x_2, ..., x_i)$ is connected to p_1 or some vertex x' in the sequence $(x_{i+1}, x_{i+2}, ..., x_k)$ is connected to p_2, or vice versa (remember that the paths connecting x and x' to p_1 and p_2 could intersect). This can be done by computing the shortest path trees in P, from both p_1 and p_2 in linear time [9]. Then we define for each $e_i = (x_i, x_{i+1})$:

1. $R_1^P(p_1, i)$ is the maximum of the shortest distances from p_1 to one of the vertices in $(x_1, x_2, ..., x_i)$, and
2. $S_1^P(p_2, i)$ is the maximum of the shortest distances from p_2 to one of the vertices in $(x_{i+1}, x_{i+2}, ..., x_k)$.
3. $R_2^P(p_2, i)$ is the maximum of the shortest distances from p_2 to one of the vertices in $(x_1, x_2, ..., x_i)$, and
4. $S_2^P(p_1, i)$ is the maximum of the shortest distances from p_1 to one of the vertices in $(x_{i+1}, x_{i+2}, ..., x_k)$.

Following Theorem 1, after $O(n)$ time preprocessing to compute the shortest path trees to x_l's, $l \in [k]$, from p_1 and p_2 [9], and computing and updating the necessary heaps (for the distances from p_1 to $(x_1, x_2, ..., x_i)$ and from p_2 to $(x_{i+1}, x_{i+2}, ..., x_k)$, or vice versa), it takes $O(\log n)$ time to compute each $R_1^P(p_1, i)$, $R_2^P(p_2, i)$, $S_1^P(p_2, i)$,$S_2^P(p_1, i)$, for $i \in [k-1]$.

Let the vertices in Q be ordered in counterclockwise order from q_1 to q_2 as $q_1(= z_0), z_1, z_2, ..., z_\ell, q_2(= z_{\ell+1})$, $\ell \in O(n)$. Symmetrically, we define for each edge $e'_j = (z_j, z_{j+1})$:

1. $T_1^Q(q_1, j)$ is the maximum of the shortest distances from q_1 to one of the vertices in $(z_1, z_2, ..., z_j)$, and
2. $U_1^Q(p_2, j)$ is the maximum of the shortest distances from q_2 to one of the vertices in $(z_{i+1}, z_{i+2}, ..., z_\ell)$.
3. $T_2^Q(q_2, j)$ is the maximum of the shortest distances from q_2 to one of the vertices in $(z_1, z_2, ..., z_j)$, and
4. $U_2^Q(q_1, j)$ is the maximum of the shortest distances from q_1 to one of the vertices in $(z_{i+1}, z_{i+2}, ..., z_\ell)$.

Similarly, after a linear time preprocessing, each $T_1^Q(p_1, j)$, $T_2^Q(p_2, j)$, $U_1^Q(p_2, j)$, $U_2^Q(q_1, j)$, for $j \in [\ell - 1]$, can be computed in $O(\log n)$ time. The optimal solution (value) for Case 1, $OPT_1(P, Q)$, can be defined as

$$OPT_1(P,Q) = \min_{i \in [k-1], j \in [\ell-1]} \{$$

$$\max\{R_1^P(p_1, i) + |p_1q_1| + T_1^Q(q_1, j), S_1^P(p_2, i) + |p_2q_2| + U_1^Q(p_2, j)\},$$

$$\max\{S_2^P(p_1,i)+|p_1q_1|+T_1^Q(q_1,j), R_2^P(p_2,i)+|p_2q_2|+U_1^Q(q_2,j)\},$$
$$\max\{R_2^P(p_2,i)+|p_2q_2|+T_2^Q(q_2,j), S_2^P(p_1,i)+|p_1q_1|+U_2^Q(p_1,j)\},$$
$$\max\{S_1^P(p_2,i)+|p_2q_2|+T_2^Q(q_2,j), R_1^P(p_1,i)+|p_1q_1|+U_2^Q(q_1,j)\}\}.$$

Note that since we need to pair 8 cases, half of them are the same; hence in the above equation we only have 4 cases. Since $i \times j \in O(n^2)$, and all the $R(-), S(-), T(-)$ and $U(-)$ items can be computed in $O(\log n)$ time. $OPT_1(P,Q)$ can be computed in $O(n^2 \log n)$ time. The lemma is proven. □

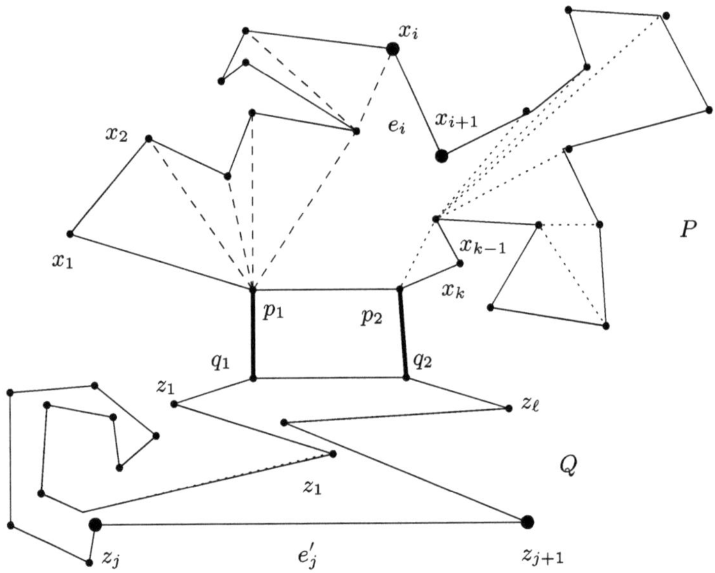

Fig. 3. Case 1 of the twin bridges problem. The dashed (resp. dotted) tree represents the shortest path tree in P from p_1 (resp. p_2).

Lemma 2. *Given two disjoint simple polygons P, Q each with at most n vertices, when the two bridges p_1q_1 and p_2q_2 are fixed, case 2 of the 2-bridge problem can be solved in $O(n^2 \log n)$ time.*

Proof. WLOG, suppose the solution in case 2 is determined by two vertices $x, y \in V(P)$ (or symmetrically when they are in $V(Q)$). If $\delta_P(x,y) > \delta_P(x,p_1) + |p_1q_1| + \delta_Q(q_1,q_2) + |q_2p_2| + \delta_P(p_2,y)$, then we can replace the distance from x to y in the new geometric structure by the latter. Following the data structure by Guibas and Hershberger [10], $\delta_P(x,p_1), \delta_Q(q_1,q_2)$ and $\delta_P(p_2,y)$ can all be computed in $O(\log n)$ time. Since we have to enumerate $O(n^2)$ pairs of x and y to select the minimum solution $OPT_2(P,Q)$, the total running time is $O(n^2 \log n)$. Therefore, the lemma is proven. □

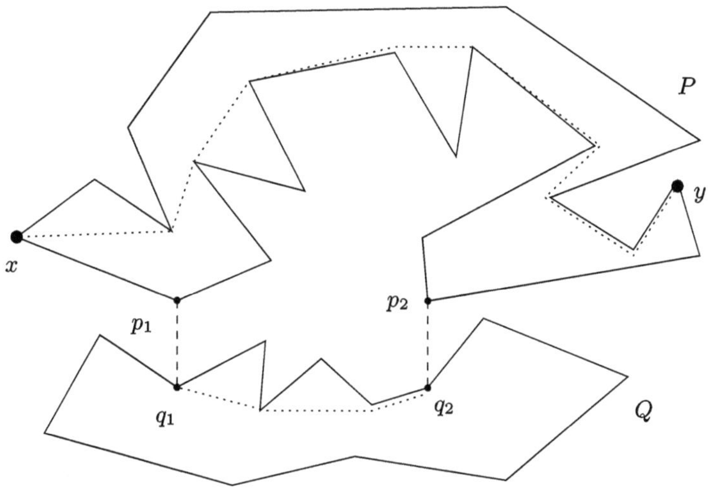

Fig. 4. Case 2 of the twin bridges problem. Note that we possibly need to enumerate all pairs of x and y in $V(P)$ (symmetrically in $V(Q)$ as well).

Theorem 2. *Given two non-intersecting simple polygons P and Q, each with at most n vertices, the twin bridges problem can be solved in* $O(n^6 \log n)$ *time.*

Proof. We could enumerate all four endpoints of the bridges $p_1, p_2 \in V(P)$ and $q_1, q_2 \in V(Q)$ in $O(n^4)$ time. We have two cases to form twin bridges: either (p_1q_1, p_2q_2) or (p_1q_2, p_2q_1). Each time after the two bridges are fixed, both Case 1 and 2 can be solved in $O(n^2 \log n)$ time following Lemma 1 and Lemma 2. Therefore, the twin bridges problem can be solved in $O(n^6 \log n)$ time. □

We have two comments here. First, we comment that even when P, Q are both convex, it does not seem that the problem can be solved faster. In fact, when P, Q are both convex, we do not have Case 2. But solving Case 1 alone still involves computing the maximum distance for something like $f(1..i)$ and $f(i+1..k)$ in the MinMax-Break problem using a Max-Heap in $O(\log n)$ time. And we do not know if convexity helps here. Second, if we require that p_i sees q_i, $i \in \{1, 2\}$, a visibility constraint can be easily added when enumerating them. Hence, the special case of the twin bridges problem that the endpoints of a bridge must be visible, i.e., p_i must see q_i (for $i \in \{1, 2\}$), can also be solved in $O(n^6 \log n)$ time.

3.2 When the Bridges Are Geodesic

When the bridges are geodesic, Lemma 1 and Lemma 2 still hold. The only difference is that the Euclidean lengths of the bridges, $|p_1q_1|$ and $|p_2q_2|$, must be replaced by the shortest path distance between p_i and q_i ($i \in \{1, 2\}$), taking P and Q as obstacles. This can be preprocessed in $O(n^2 \log n)$ time as follows. We compute the shortest path tree from a single vertex $p \in V(P)$ (and $q \in V(Q)$) to all other vertices of P and Q (taking P, Q as obstacles) in $O(n \log n)$ time, using the algorithm by Hershberger and Suri [11]. Repeating this on $|P|+|Q|$ trees and storing all the distances from the root to all the corresponding leaves, we could return the shortest path distance between $p \in V(P)$ and $q \in V(Q)$ in constant time—after a total of $O(n^2 \log n)$ time preprocessing. Then the remaining parts of the algorithms would be very much the same, and we only need to change the Euclidean distance between p_i and q_i to their corresponding geodesic distances. We state the following results without giving further details.

Corollary 1. *The twin bridges problem can be solved in* $O(n^6 \log n)$ *time even when the two bridges are geodesic.*

4 NP-Hardness for Connecting Multiple Polygons with Multiple Twin Bridges

In this section, we prove that the following generalized version of the twin bridges problem in multiple polygons (TBMP) is NP-hard: Given a sequence of disjoint simple polygons $P_1, P_2, ..., P_m$, each with $O(n)$ vertices, add two bridges between P_i and P_{i+1}, for $i \in [m-1]$, such that the constrained diameter of the resulting geometric structure is minimized. (The structure is a polygon with $m-1$ holes when no two bridges intersect.) See Fig. 5 and Fig. 6 for an illustration.

We reduce Set-Partition for Rational Pairs (SPRP), which was shown NP-hard by Wojtczak [16]. Formally, SPRP is defined as follows: given m pairs of rational numbers (a_i, b_i), $i \in [m]$, such that $\sum_i (a_i + b_i) = 2A$, partition each pair of rationals (a_i, b_i) into two sets S_1 and S_2 respectively such that $\sum_{x \in S_1} x = \sum_{y \in S_2} y = A$.

The polygon P_i's are constructed as follows. We have $m-1$ skinny rectangle R_i's with height $2H$ evenly distributed, and vertically centered at line $X = 0$ (dashed in Fig. 5 and Fig. 6). Then for each rectangle R_i, we attach two narrow polygonal tubes starting from $v_{i,1}$ and $v_{i,2}$ with a shortest path distance a_i

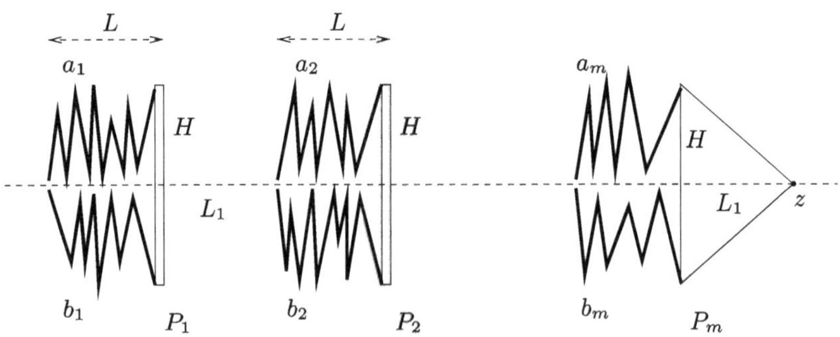

Fig. 5. The NP-hardness proof for TBMP: the general structure.

and b_i respectively to the top-right and bottom-right vertices (v_{i,n_1}, v_{i,n_2} of R_i respectively. The horizontal distance from $v_{i,1}$ and $v_{i,2}$ to the right edge of R_i is L; moreover, the distance between $v_{i,1}$ and $v_{i,2}$ to any vertex in R_{i-1} is the smallest among all other vertices in P_i. Then we have the resulting polygon P_i for $i \in [m-1]$. Note that $v_{i,1}, v_{i,2}$ are arbitrarily close, and the horizontal distance from them to the right edge of P_{i-1} is L_1; hence, the distance from them to v_{i-1,n_1} and v_{i-1,n_2} is $\sqrt{H^2+L_1^2}$ for $i \in [2,m]$. Moreover, for the attached skinny polygonal tubes, each (vertical) segment can be of a distance $a_i/O(n)$ or $b_i/O(n)$—depending on if they are above the dashed line $X=0$, hence it takes $O(n)$ time to construct P_i, for $i \in [1, m-1]$. For P_m the right part is constructed slightly differently: we have a triangle whose (vertical) base edge is of length $2H$, the height from the third vertex z to this base edge is L_1 and the two non-vertical edges are of length $C = \sqrt{H^2+L_1^2}$.

Therefore, it takes $O(mn)$ time to construct all P_i's, for $i \in [m]$, where $O(n)$ is the size of each P_i.

Note that the distance from $v_{i,1}$ and $v_{i,2}$ to $v_{i-1.n_1}$ and v_{i-1,n_2}, for $i \in [2,m]$, can be considered as a constant C—this can be done by reducing the width of R_i's to arbitrarily small. Although when we connect $v_{i,2}$ to v_{i-1,n_1} and $v_{i,1}$ to v_{i-1,n_2} as bridges, they could intersect at the right ends (which we allow as we have claimed earlier). (See Fig. 6.)

We claim that SPRP has a solution if and only if $2(m-1)$ bridges can be inserted between P_i and P_{i+1}, $i \in [m-1]$ such that the constrained diameter of the resulting geometric structure is $A+mC$.

The "only-if" part is obvious, so we focus on the "if" part. If $2(m-1)$ bridges can be inserted between P_i and P_{i+1}, $i \in [m-1]$ such that the constrained diameter of the resulting geometric structure is $A+mC$, the first thing we note that is the bridges between P_i and P_{i+1} must be between their vertices. Then, by construction, we must use the pairs between $\{v_{i,1}, v_{i,2}\}$ and $\{v_{i-1,n_1}, v_{i-1,n_2}$, which has a distance C. Since in P_m we also need a distance C to connect z to either of the two added polygonal tubes, we would need a total distance of mC to connect $v_{1,1}$ and $v_{1,2}$ to C, not counting the added polygonal tubes in P_j's, $j \in [m]$.

Then, clearly, since the total constrained diameter is of length $A+mC$, we must connect m added tubes from $v_{1,1}$ and $v_{1,2}$ to z. This gives a partition of all the (a_i, b_i) pairs, for $i \in m$. More precisely, if a tube in P_j (of length a_j or b_j) is on the path from $v_{1,1}$ to Z, we put it in S_1; otherwise we put it in S_2.

We then have the following theorem.

Theorem 3. *The twin bridges problem in multiple polygons is NP-hard.*

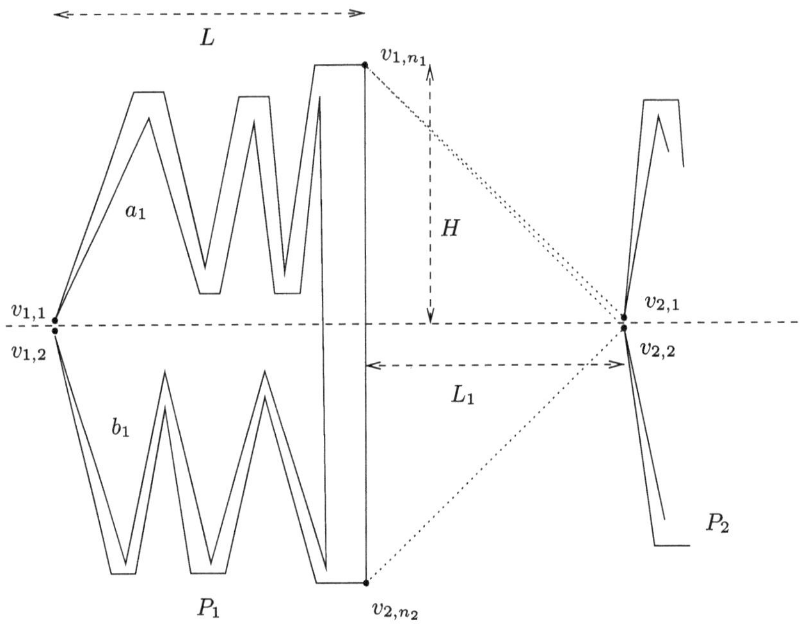

Fig. 6. The NP-hardness proof for TBMP: the amplified view at P_1 and P_2.

5 Concluding Remarks

In this paper, we consider the twin bridges problem in polygons. For two polygons we give an $O(n^6 \log n)$ time algorithm and for multiple polygons we prove that it is NP-hard. It is interesting that even though this is a geometric problem, the geometric property (in fact, even convexity) does not seem to help much. It would be interesting to see what property or constraint we could find or impose to solve the problem faster at least for the two-polygon case.

Acknowledgments. This research is partially supported by NSF under project CNS-2243010; by the National Key R&D Program of China (project 2020YFB1406700), and also by NSFC of China (project 62272279). LW is supported by the GRF grant from Hong Kong SAR, China, under project CityU 11218423.

References

1. Ahn, H., Cheong, O., Shin, C.: Building bridges between convex regions. Comput. Geom. **25**(1–2), 161–170 (2003)
2. Bhattacharya, B.K., Benkoczi, R.: On computing the optimal bridge between two convex polygons. Inf. Process. Lett. **79**(5), 215–221 (2001)
3. Bhosle, A.M., Gonzalez, T.F.: Exact and approximation algorithms for finding an optimal bridge connecting two simple polygons. Int. J. Comput. Geometry Appl. **15**(6), 609–630 (2005)

4. Bhosle, A.M., Gonzalez, T.F.: Finding optimal geodesic bridges between two simple polygons. In: Proceedings of the 23rd Annual Canadian Conference on Computational Geometry, CCCG. pp. 1–6 (2011)
5. Cai, L., Xu, Y., Zhu, B.: Computing the optimal bridge between two convex polygons. Inf. Process. Lett. **69**(3), 127–130 (1999)
6. Cormen, T.H., Leiserson, C.E., Rivest, R.L., Stein, C.: Introduction to Algorithms, 3rd Edition. MIT Press (2009)
7. Feng, Z., Fernau, H., Zhu, B.: Optimal bridge, twin bridges and beyond: Inserting edges into a road network to minimize the constrained diameters. In: Ghosh, S., Zhang, Z. (eds.) Algorithmic Aspects in Information and Management - AAIM 2024 - 18th Annual International Conference, Dallas, Texas, September 21-23, 2024. Proceedings. Lecture Notes in Computer Science, vol. 15179, pp. 94–108. Springer (2024)
8. Gudmundsson, J., Wong, S.: Improving the dilation of a metric graph by adding edges. ACM Trans. Algorith. **18**(3), 20:1–20:20 (2022)
9. Guibas, L., Hershberger, J., Leven, D., Sharir, M., Tarjan, R.E.: Linear-time algorithms for visibility and shortest path problems inside triangulated simple polygons. Algorithmica **2**(1), 209–233 (1987)
10. Guibas, L.J., Hershberger, J.: Optimal shortest path queries in a simple polygon. J. Comput. Syst. Sci. **39**(2), 126–152 (1989)
11. Hershberger, J., Suri, S.: An optimal algorithm for euclidean shortest paths in the plane. SIAM J. Comput. **28**(6), 2215–2256 (1999)
12. Kim, S.K., Shin, C.: Computing the optimal bridge between two polygons. Theor. Comput. Syst. **34**(4), 337–352 (2001)
13. Tan, X.: On optimal bridges between two convex regions. Inf. Process. Lett. **76**(4–6), 163–168 (2000)
14. Tan, X.: Finding an optimal bridge between two polygons. Int. J. Comput. Geometry Appl. **12**(3), 249–262 (2002)
15. Wang, D.P.: An optimal algorithm for constructing an optimal bridge between two simple rectilinear polygons. Inf. Process. Lett. **79**(5), 229–236 (2001)
16. Wojtczak, D.: On Strong NP-Completeness of Rational Problems. In: Fomin, F.V., Podolskii, V.V. (eds.) CSR 2018. LNCS, vol. 10846, pp. 308–320. Springer, Cham (2018). https://doi.org/10.1007/978-3-319-90530-3_26

A Maximal Independent Set Heuristic for the Stochastic Critical Node Detection Problem

Tuguldur Bayarsaikhan[1], Altannar Chinchuluun[2](✉), and Ashwin Arulselvan[3]

[1] Department of Information and Computer Science, National University of Mongolia, Ulaanbaatar, Mongolia
18B1NUM0267@stud.num.edu.mn

[2] Business School, National University of Mongolia, Ulaanbaatar, Mongolia
altannar@num.edu.mn

[3] Department of Management Science, University of Strathclyde, Glasgow, UK
ashwin.arulselvan@strath.ac.uk

Abstract. Critical node detection involves identifying key nodes whose removal significantly disrupts network connectivity or performance. The evaluation of vulnerability in networks is crucial because many real-world problems are represented as graphs with uncertainty, including social networks, communication infrastructures, epidemiological models, and transportation systems. In this paper, we consider a stochastic Critical Node Detection Problem (SCNDP) with edge uncertainty, aiming to minimize the expected pairwise connectivity (EPC) in the resulting residual network. We propose a heuristic method for the SCNDP and compare it with the existing algorithm. Experimental results performed on random graphs with different edge-probability configurations demonstrate the effectiveness of the heuristics.

Keywords: Critical node detection · Integer programming · Stochastic programming

1 Introduction

Problems in real life can be effectively modeled as networks due to their flexibility and broad range. Networks serve as an essential tool in various research domains, such as social network analysis [18], traffic systems [11], and epidemic transmission networks [6]. Network models have been used to investigate the spread of epidemics, such as COVID-19, throughout a community and to identify key individuals whose infection would significantly increase further infections, which helps prevent disease transmission [6].

The critical node detection problem was first introduced by [2]. Earlier formulations encompass Mixed-Integer Linear Programming (MILP) methods, which encounter computational challenges as the graph size increases. Consequently,

A. Chinchuluun et al. (Eds.): AAIM 2025, LNCS 16379, pp. 14–25, 2026.
https://doi.org/10.1007/978-981-95-5657-1_2

the advancement of heuristic and metaheuristic methods provides effective solutions for larger networks but lacks solution quality. The variants of the problem have expanded by integrating stochastic versions [8,12] and multi-level defender-attacker scenarios, which analyze networks across multiple attack stages [3].

Although many studies related to the CNDP were conducted [1], only a few of them considered the assessment of networks with uncertainty. In real life, those uncertainties occur in most cases due to the dynamic nature of the network, environmental variability, model and parameter uncertainty, and inaccuracies in data collection or data processing. Such uncertainties considerably affect critical elements of the network, leading to potential misjudgments of network vulnerability if ignored [5,15,16].

[12] studied the stochastic version of the SCNDP over trees, where node attacks might fail probabilistically. They showed that this problem generalizes to the Critical Element Detection Problem (CEDP), where there is a chance that both nodes and edges will not be removed. Even in simplified settings, the authors establish that the stochastic version of the problem is NP-complete. The deterministic counterpart of the problem over trees with connection specified cost for each pair of distinct nodes is still NP-complete as demonstrated by [7]. This highlights the inherent complexity introduced by stochastic uncertainty. Both linear and nonlinear Mixed-Integer Linear Programming formulations were proposed and solved as an exact method utilizing Benders decomposition that effectively works for large instances.

[8] considered the CNDP with probabilistic graphs and constructed an efficient *Fully Polynomial-Time Randomized Approximation Scheme* (FPRAS) to compute the EPC. Unlike deterministic graphs, where the topology is fixed, they treated the original graph as a generative probabilistic model producing numerous graph realizations, each edge associated with a probability. The goal is to minimize the EPC, and computing EPC exactly is $\#P$-complete because of exponentially many graph realizations with 2^m different scenarios to analyze, where m is the number of edges in the input graph. In addition, the MILP model they introduced also suffers from having exponentially many variables and constraints corresponding to the scenarios. Hence, they proposed a heuristic approach named Rounding the Expected Graph Algorithm (REGA), which solves a reduced MILP with an aggregated set of variables and constraints by leveraging an FPRAS to estimate EPC efficiently without evaluating all possible realizations. A CNDP variant with nondeterministic connection costs was formulated as a robust optimization problem by [10].

In this work we adapt the MIS based heuristic introduced in [2] to the Stochastic version and we implemented a local search heuristics to enhance the solution quality. The rest of the article is structured as follows. In Sect. 2, we review the stochastic CNDP problem providing two alternative definitions and demonstrating the equivalence of their objective functions. Section 3 describes the description of the proposed algorithm. In Sect. 4, gives parameter settings and descriptions of datasets, experiment settings and experimental results. Finally, Sect. 5 summarizes the work and potential research directions.

2 Problem Formulation

Given a graph $G(V, E)$ with a set of vertices V, a set of edges E, and a budget K, the CNDP selects a set of K nodes, S, to attack or remove so that the pairwise connectivity of the remaining nodes in $V \setminus S$ is minimized. Let $G_{V \setminus S}$ be the node-induced subgraph with nodes $V \setminus S$.

There are stochastic CNDP papers in the literature [8,12]. It assumes that the outcome of attacks on nodes is uncertain, i.e., there are certain probabilities that the nodes to be successfully removed if attacked. [8] considered the CNDP where edges have certain probabilities to exist.

Definition 1 (Stochastic CNDP with uncertain edges [8]). *Given a graph $G(V, E)$ with a set of vertices V, a set of edges E, and a budget K. Let $\pi : E \rightarrow (0, 1]$ be a predefined function associated with edges. Let $G'(V, E')$ be a random graph obtained from G by removing each edge $e \in E$ with probability $1 - \pi(e)$. Then, the stochastic CNDP with uncertain edges selects a set of K nodes, S, to attack or remove so that the EPC of the remaining nodes in the node-induced subgraph of G' is minimized.*

Let $\mathcal{L}_G(i, j)$ be the set of all distinct simple paths (without cycles) between $(i, j) \in V \times V$, $i \neq j$, in G. Let $l = \{e_1, e_2, ..., e_q\} \in \mathcal{L}_G(i, j)$ be a simple path between i and j, where i is the starting point of e_1 and j is the end point of e_q. Then, the probability that the nodes i and j are connected in G' through the path $l(i, j)$ is $\prod_{e \in l} \pi(e)$. The probability of overall connectivity between any pair of nodes i and j in G' is given by the following formula:

$$P(\bigcup_{l \in \mathcal{L}_G(i,j)} \{l \text{exists in } G'\}) = \sum_{Q \subseteq \mathcal{L}_G(i,j)} (-1)^{|Q|+1} \prod_{e \in \{\text{any path in } Q\}} \pi(e).$$

Our problem can be formulated as the following optimization problem:

$$\begin{aligned} \min_{S \subset V} \quad & \sum_{i \neq j \in V \setminus S} \sum_{Q \subseteq \mathcal{L}_G(i,j)} (-1)^{|Q|+1} \prod_{e \in \{\text{any path in } Q\}} \pi(e). \\ \text{s.t.} \quad & |S| \leq K. \end{aligned}$$

For large graphs, this formulation is very difficult to implement. Therefore, we need another formulation.

We borrow the concept of live-edge graphs introduced by [13] for the influence maximization problem. Let Ω be the set of all possible scenarios of the stochastic graph G. We obtain a random scenario $w \in \Omega$ by randomly blocking or activating (to be live) each edge in G, resulting in a static subgraph of $G^w = (V, E^w)$, called a live-edge graph.

Definition 2 (Live-edge graph). *We select each edge $(i, j) \in E$ independently with probability $\pi(i, j)$. The selected edge is declared to be "live" and all other edges are declared to be "blocked". Let us denote the set of live edges by E^w and* *the resulting graph by $G^w(V, E^w)$ or G^w in short. Then each edge $(i, j) \in E^w$ indicates that there is an edge between i and j with a probability of 1 in scenario w.*

Therefore, the pairwise connectivity of the stochastic CNDP problem can be calculated based on the live-edge graph scenarios. In scenario w, let $\sigma^w(S)$ be the pairwise connectivity of G^w after deleting the node set S, i.e., $\sigma^w(S)$ is the pairwise connectivity of the node-induced subgraph $G^w[V \backslash S]$. Then the EPC function can be calculated by the following formula:

$$\sigma(S) = \sum_{w \in \Omega} p^w \, \sigma^w(S) \,.$$

Here, the probability of a scenario w is $p^w = \prod_{(i,j) \in E^w} \pi_{ij} \prod_{(i,j) \in E \backslash E^w} (1 - \pi_{ij})$, and the number of all possible live-edge graphs or scenarios is 2^m, where $m = |E|$.

Subsequently, our problem can be formulated as the following optimization problem.

$$\begin{aligned} \min_{S \subset V} \; & \sum_{w \in \Omega} p^w \, \sigma^w(S) \\ \text{s.t.} \; & |S| \leq K. \end{aligned}$$

One way to model this is by extending the deterministic formulation introduced in [2] to a two-stage stochastic program. We introduce the 0–1 variables s_i for each $i \in V$ indicating whether node i is chosen as critical or not and for each scenario $w \in \Omega$, we introduce 0–1 connectivity variables u_{ij}^w that model whether nodes i and j are connected.

$$(MIP_F) \qquad \begin{aligned} \min_{S \subset V} \; & \sum_{w \in \Omega} p^w \sum_{i,j \in V} u_{ij}^w \\ \text{s.t.} \; & \sum_{i \in V} s_i \leq K \\ & u_{ij}^w + s_i + s_j \geq 1, \; \forall w \in \Omega, \; \forall (i,j) \in E^w, \\ & u_{ij}^w + u_{j,k}^w - u_{ki}^w \leq 1, \; \forall (i,j,k) \in V, \\ & u_{ij}^w - u_{j,k}^w + u_{ki}^w \leq 1, \; \forall (i,j,k) \in V, \\ & -u_{ij}^w + u_{j,k}^w + u_{ki}^w \leq 1, \; \forall (i,j,k) \in V, \\ & s_i \in \{0,1\}, \; \forall i \in V, \\ & u_{i,j}^w \in [0,1], \; \forall w \in \Omega, \; \forall i \in V. \end{aligned}$$

3 Methodology

In this section, we present heuristic approaches to the SCNDP. Each approach is explained in detail, and the descriptions of the algorithms are provided.

3.1 Benchmark Methods

To assess the performance of our proposed algorithms, we implemented the Rounding the Expected Graph Algorithm (REGA) and the Component Sampling Procedure (CSR) for EPC estimation as described by [8].

Component Sampling Procedure. The component sampling procedure (CSP) efficiently estimates EPC by sampling graph components instead of entire graphs. It has the advantage of having polynomial-time complexity due to an FPRAS, with the running time bounded by a polynomial in $1/\epsilon, log(1/\delta)$, and the input size, and it significantly reduces computational overhead compared to naive Monte Carlo methods. CSP employs an importance sampling strategy, selecting nodes uniformly and performing localized breadth-first searches to estimate EPC values quickly. To detect critical nodes in large graphs efficiently, we parallelized the computation. We provide the description of CSP in Algorithm 1 for calculating $\sigma(S)$ as described in [8].

Algorithm 1. (ε, δ) Component Sampling Procedure (CSP) for calculating $\sigma(S)$

Input: Graph $G = (V, E)$ with edge survival probs p_{ij}, accuracy (ε, δ)
Output: Estimator $\widehat{\mathcal{E}}$

$P_E \leftarrow \sum_{e \in E} p_e$
if $P_E < \frac{\varepsilon}{2} n^{-2}$ **then**
    return $\widehat{\mathcal{E}} \leftarrow P_E$
end if
$C_2 \leftarrow 0$
for $i = 1 \ldots N(\varepsilon, \delta)$ **do**
    Select node $u \in V$ uniformly at random
    Run BFS from u; when exploring edge (v, w), keep it w.p. p_{vw} (else discard)
    Let S_i be number of visited nodes (including u)
    $C_2 \leftarrow C_2 + (S_i - 1)$
end for
$\widehat{\mathcal{E}} \leftarrow \dfrac{n\, C_2}{2N(\varepsilon, \delta)}$
return $\widehat{\mathcal{E}}$

We take $N(\epsilon, \delta) \geq 4(e-2) \ln\left(\frac{1}{\delta}\right) \frac{1}{\epsilon^2 EPC}$ in the above algorithm. A lower bound on EPC is taken to calculate $N(\epsilon, \delta)$, in order to ensure that $\widehat{\mathcal{E}}$ is a (ϵ, δ)-estimator of EPC, i.e., we have

$$\Pr\left[(1-\epsilon)\text{EPC} \leq \widehat{\mathcal{E}} \leq (1+\epsilon)\text{EPC} \geq (1-\delta)\right]$$

REGA Algorithm. REGA is a heuristic algorithm designed to efficiently solve SCNDP. It first replaces the huge twostage mixed integer program MIP_F with a much smaller linear program by averaging the scenario constraints using their probabilities. After solving this relaxed model, it rounds the fractional solution to choose the K nodes to delete. A local search procedure is applied after deleting K nodes, where nodes in the current deletion set are iteratively swapped with remaining nodes if the swap reduces EPC. We provide the reduced MIP formulation, MIP_R, and the description of REGA in Algorithm 2 as described in [8], below.

Reduced Mixed Integer Program, MIP_R :

$$
\begin{aligned}
\min\ & \sum_{i<j}(1 - x_{ij}) \\
\text{s.t. } & \sum_{i=1}^{n} s_i \le K \\
& x_{ij} \le s_i + s_j + 1 - p_{ij},\ (i,j) \in E, \\
& x_{ij} + x_{jk} \ge x_{ik},\ (i,j) \in E,\ k = 1,\dots,n, \\
& s_i \in \{0,1\},\ x_{ij} \in [0,1],\ i,j = 1,\dots,n
\end{aligned}
$$

Algorithm 2. Rounding the Expected Graph Algorithm (REGA)

Input: Graph $G = (V, E)$, budget k
Output: Deletion set D

$D \leftarrow \emptyset$
for $t = 1, \dots, k$ **do**
  Solve the LP relaxation of MIP_R to obtain fractional solution s
  $u \leftarrow \arg\max_{i \in V \setminus D} s_i$
  $D \leftarrow D \cup \{u\}$
  $s_u \leftarrow 1$ in MIP_R
end for
repeat
  improved $\leftarrow$ **false**
  for all $(u, v) \in D \times D$ **do**
    Estimate $\sigma(D - \{u\} + \{v\})$
    if $\sigma(D - \{u\} + \{v\}) < \sigma(D)$ **then**
      $D \leftarrow D - \{u\} + \{v\}$
      improved $\leftarrow$ **true**
    end if
  end for
until *improved* $=$ **false**
return D

3.2 Proposed Method

Greedy Heuristic with MIS. The greedy algorithm with a maximal independent set (MIS) is adapted from the heuristic algorithm of [2], which is originally a deterministic algorithm. We extend it by incorporating the CSR, an EPC estimation algorithm, to make it suitable for stochastic networks. Since MIS generation is inherently random, multiple MIS samples are generated, and the set yielding the lowest EPC is chosen. To improve robustness, the algorithm is executed a certain number of times, and we integrated the Cost-Effective Lazy Forward (CELF) algorithm introduced by [14] for the influence maximization problem. CELF avoids recalculating the EPC for all nodes in each iteration. In the first iteration, it calculates the EPC for all nodes and sorts them. Then, it only recalculates the spread of the top-ranked node in each subsequent iteration. This significantly speed up the process without compromising the quality of the solutions found by the algorithm. The performance is summarized in Fig. 1. Finally, the best initial solution set is refined using the same local search procedure in REGA.

Algorithm 3. Greedy heuristic with MIS

Input: $G(V, E)$, K, $\pi : V \to (0, 1]$
Output: S

$MIS \leftarrow MaximalIndepSet(G)$
for $i = 1, \ldots, |V| - K$ **do**
  $v \leftarrow \arg\min_{j \in V \setminus MIS} \sigma(V \setminus \{MIS \cup j\}) - \sigma(V \setminus MIS)$
  $MIS \leftarrow MIS \cup v$
end for
$S \leftarrow V \setminus MIS$

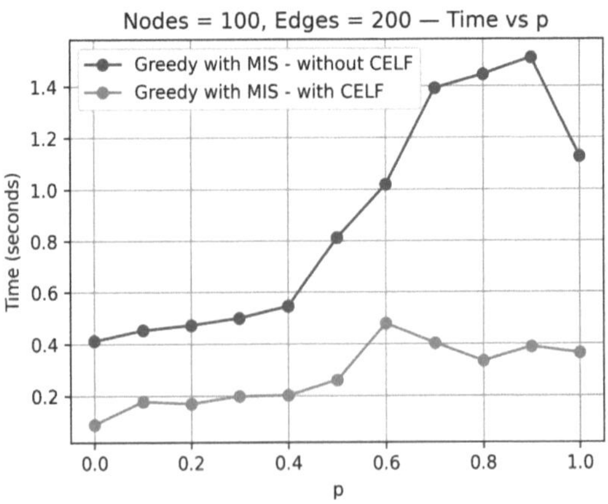

Fig. 1. Runtime of Greedy with MIS—with and without CELF optimization

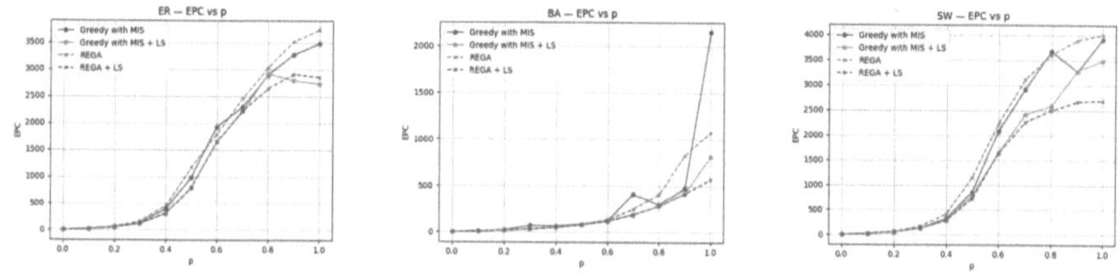

(a) Erdos–Renyi Network (b) Barabási–Albert Network (c) Small-world Network

Fig. 2. Comparing performance of the algorithms: EPC vs p

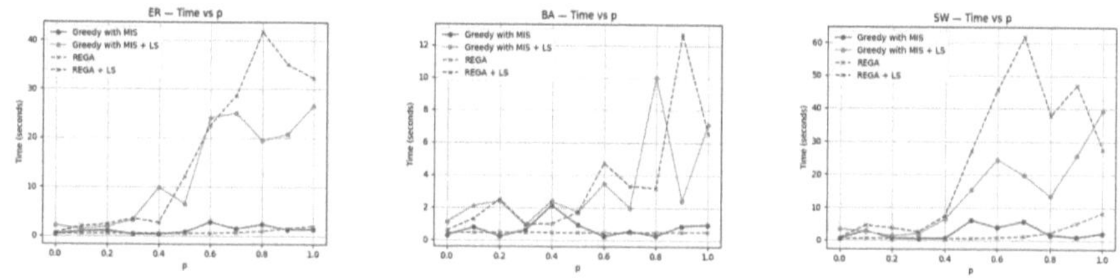

(a) Erdos–Renyi Network (b) Barabási–Albert Network (c) Small-world Network

Fig. 3. Comparing performance of the algorithms: Runtime vs p

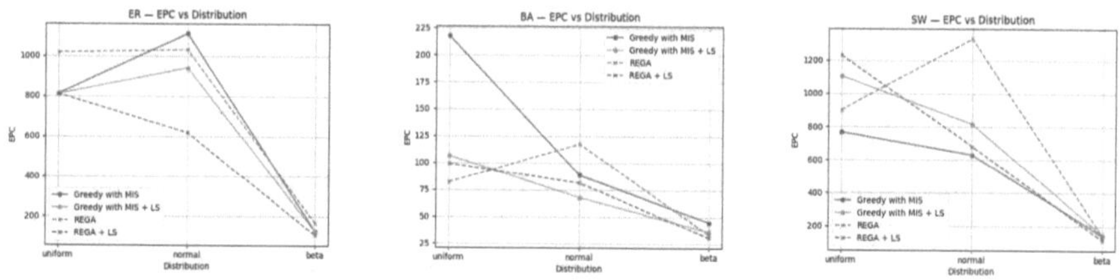

(a) Erdos–Renyi Network (b) Barabási–Albert Network (c) Small-world Network

Fig. 4. Comparing performance of the algorithms (heterogeneous distributions): EPC vs distribution

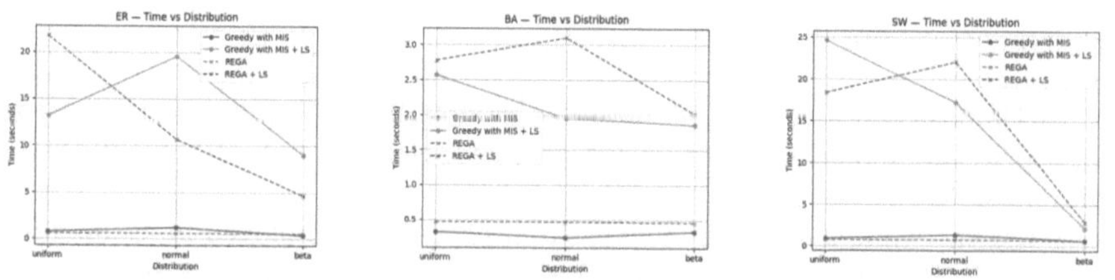

(a) Erdos–Renyi Network (b) Barabási–Albert Network (c) Small-world Network

Fig. 5. Comparing running time of the algorithms (heterogeneous distributions): Runtime vs distribution

4 Experiments

We demonstrate through experiments the effectiveness of our proposed algorithms in comparison to the REGA heuristic. We evaluated our approaches using randomly generated synthetic graphs with customized edge probability values or distributions for our experimental settings.

4.1 Edge-Probability Settings

We evaluate each proposed method under the following probability settings for edge uncertainty:

1. *Uniform probability for all edges*: $\pi(e) = p$ for all $e \in E$, following Dinh and Thai [8].
2. *Heterogeneous probability distributions for all edges:*

$$\pi(e) \sim \begin{cases} \text{Uniform}(0,1) \\ \text{Beta}(2,5) \\ \text{Normal}(0.5, 0.2) \text{ truncated to } (0,1] \end{cases} \qquad \text{for all } e \in E.$$

4.2 Dataset

To evaluate our methods more comprehensively, we consider a variety of graph instances. We adopt most of the parameters from [8] as described below:

1. *Erdős–Rényi (ER)*: A random graph of 100 nodes and 200 edges, where each edge is uniformly distributed at random [9].
2. *Barabási–Albert (BA)*: A scale-free network of 100 nodes and 200 edges, generated by the preferential attachment mechanism, yields a distribution of degrees of power law [4].
3. *Watts–Strogatz (WS)*: A small world network with 100 nodes, derived from a two-dimensional lattice with a rewiring probability of 0.3 [17].

4.3 Heuristic Settings

Due to the stochastic nature of finding the maximal independent set (MIS), we perform 40 independent trials for graphs, selecting the MIS yielding the minimum EPC.

We use the same 2-exchange local search procedure employed in [8] for both the heuristic algorithm and REGA. The local search looks for better solutions by changing one node at a time in the current set and checking to see if the goal is met. We used 10,000 samples for each local search iteration for computational efficiency and low variance in the results. However, 100,000 samples are used to estimate the final EPC, ensuring a highly accurate estimate.

4.4 Environment

The algorithms were implemented in Python on a 64-bit Ubuntu platform featuring an Intel i7 3.8 GHz processor and 64 GB of memory. Linear programming problems were solved using the *SciPy* optimization library.

4.5 Experimental Results

Across algorithms, lower EPC values and shorter runtime indicate the best performance. Figures 2 and 3 illustrate the experimental results obtained with same uniform probabilities for all edges. The naive REGA algorithm, which does not incorporate a local search procedure, produced lower-quality solutions and had low efficiency. However, solution quality improved significantly when integrated with local search but resulted in a considerably increased runtime. In comparison, the greedy MIS heuristic outperforms the naive REGA. When local search was included with greedy MIS, solution quality closely approached to the REGA with local search, even outperforming it in several cases, while maintaining significantly lower runtime.

Figures 4 and 5 show results for graphs with heterogeneous edge probabilities. The greedy MIS heuristic outperforms REGA in several cases. When both REGA and greedy MIS algorithms were integrated with the local search procedure, they consistently yielded the best reductions in EPC. However, greedy MIS typically exhibited faster execution.

Overall, our proposed algorithm, greedy with MIS heuristic, offers comparable solution quality to REGA, both with and without the local search procedure. Therefore, it significantly reduces computational cost, making it particularly suitable for larger graph instances.

5 Conclusion

Optimizing networks with uncertain edges presents significant challenges. We considered the CNDP with edge uncertainty and proposed the greedy with MIS heuristic method for the problem. A local search technique was used in order to improve the method. When testing the proposed methods on benchmark graphs featuring both uniform edge probabilities and heterogeneous edge probabilities, and compared the proposed method with the results obtained by applying the existing algorithm called REGA. The experimental results show that the method achieves equal or lower expected pairwise connectivity in certain cases and demonstrates faster runtime compared to REGA. Future research will involve integrating learning-based techniques and assessing the algorithm's performance on much larger graphs.

References

1. Arulselvan, A., Chinchuluun, A.: Identifying critical nodes in a network. In: Balomenos, K.P., Fytopoulos, A., Pardalos, P.M. (eds.) Handbook for Management of Threats: Security and Defense, Resilience and Optimal Strategies, pp. 325–339. Springer International Publishing (2023)
2. Arulselvan, A., Commander, C.W., Elefteriadou, L., Pardalos, P.M.: Detecting critical nodes in sparse graphs. Comput. Oper. Res. **36**, 2193–2200 (2009)
3. Baggio, A., Carvalho, M., Lodi, A., Tramontani, A.: Multilevel approaches for the critical node problem. Oper. Res. **69**(2), 486–508 (2021)
4. Barabási, A.L., Albert, R., Jeong, H.: Scale-free characteristics of random networks: The topology of the world-wide web. Physica A (2000)
5. Bernstein, A., Bienstock, D., Hay, D., Uzunoglu, M., Zussman, G.: Power grid vulnerability to geographically correlated failures: Analysis and control implications. In: Proceedings of the IEEE INFOCOM, pp. 2634–2642 (2014)
6. Chaharborj, S.S., Nabi, K.N., Feng, K.L., Chaharborj, S.S., Phang, P.S.: Controlling covid-19 transmission with isolation of influential nodes. Chaos, Solitons Fractals **159**, 112035 (2022)
7. Di Summa, M., Grosso, A., Locatelli, M.: Complexity of the critical node problem over trees. Comput. Oper. Res. **38**(12), 1766–1774 (2011)
8. Dinh, T.N., Thai, M.T.: Assessing attack vulnerability in networks with uncertainty. In: IEEE Conference on Computer Communications (INFOCOM), pp. 2380–2388 (2015)
9. Erdős, P., Rényi, A.: On the evolution of random graphs. Publ. Math. Inst. Hung. Acad. Sci. **5**, 17–61 (1960)
10. Fan, N., Pardalos, P.M.: Robust optimization of graph partitioning and critical node detection in analyzing networks. In: Wu, W., Daescu, O. (eds.) Combinatorial Optimization and Applications, pp. 170–183. Springer, Berlin Heidelberg, Berlin, Heidelberg (2010)
11. Ghosh, S., Banerjee, A., Sharma, N., Agarwal, S., Ganguly, N., Mukherjee, A.: Statistical analysis of the Indian railway network: a complex network approach. Acta Phys. Pol. B Proc. Suppl. **4**(2), 123–137 (2011)
12. Hosteins, P., Scatamacchia, R.: The stochastic critical node problem over trees. Networks **76**(3), 381–401 (2020)
13. Kempe, D., Kleinberg, J.M., Tardos, É.: Maximizing the spread of influence through a social network. In: Proceedings of the 9th International Conference on Knowledge Discovery and Data Mining (KDD 2003), pp. 137–146 (2003)
14. Leskovec, J., Krause, A., Guestrin, C., Faloutsos, C., VanBriesen, J., Glance, N.: Cost-effective outbreak detection in networks. In: Proceedings of the 13th ACM SIGKDD International Conference on Knowledge Discovery and Data Mining, pp. 420–429 (2007)
15. Lindgren, A., Doria, A., Schelén, O.: Probabilistic routing in intermittently connected networks. In: Proceedings of the ACM SIGMOBILE MCCR, pp. 19–20 (2003)
16. Soltani-Sobh, A., Heaslip, K., Stevanovic, A., El Khoury, J., Song, Z.: Evaluation of transportation network reliability during unexpected events with multiple uncertainties. Int. J. Disast. Risk Reduct. **17**, 128–136 (2016)

17. Watts, D.J., Strogatz, S.H.: Collective dynamics of small-world networks. Nature **393**(6684) (1998)
18. Weng, J., Lim, E.P., Jiang, J., He, Q.: TwitterRank: finding topic-sensitive influential twitterers. In: Proceedings of the 3rd ACM Int. Conference on Web Search and Data Mining (WSDM 2010), pp. 261–270. ACM, New York, USA (2010)

Utilizing Large Language Models for Finding All Roots to Nonlinear Systems of Equations: Solutions, Accuracy, and Prompt Design

Alexander Semenov[1], Michael J. Hirsch[2,3](✉), and Panos M. Pardalos[3,4]

[1] Department of Industrial and Management Systems Engineering, University of South Florida, Tampa, FL, USA
asemenov@usf.edu

[2] ISEA TEK, LLC, Maitland, FL, USA
mhirsch@iseatek.com

[3] TOXEUS Systems, LLC, Maitland, FL, USA
pardalos@ise.ufl.edu

[4] Department of Industrial and Systems Engineering, University of Florida, Gainesville, FL, USA

Abstract. Finding all solutions to a system of non-linear equations is a longstanding and challenging problem with applications across science and engineering. The recent emergence of large language models raises the question of whether these models can assist, or even autonomously solve, such problems. This paper reviews classical and global optimization approaches for finding all roots to systems of nonlinear equations and empirically evaluates nine contemporary large language models on a suite of small-sized representative systems. The solution quality is analyzed, common failure modes are investigated, and the impact of prompt design is considered, offering guidelines for practitioners and directions for future research.

Keywords: Systems of nonlinear equations · Large Language Models · Root finding algorithms

1 Introduction

Systems of non-linear equations (SNEs) arise in diverse domains such as robotics, chemical engineering, power systems, economics, and many others [3,4]. Depending on the characteristics of the SNEs, determining even just one solution can be NP-hard [9]. Finding all solutions to an SNE is even more complex. Solving SNEs are intimately related to Hilbert's Tenth Problem [6].

Classical numerical root-finding methods [10] (e.g., Newton's Method [2] and Broyden's Method [1], the two-most common) rely on local derivative information and may converge only to a root close to the starting point. There has been

A. Chinchuluun et al. (Eds.): AAIM 2025, LNCS 16379, pp. 26–34, 2026.
https://doi.org/10.1007/978-981-95-5657-1_3

research into reformulating an SNE as a global optimization problem [11], and approaches to find all solutions to an SNE through the use of adaptive penalty and/or deflation techniques [7,13]. These approaches normally require sophisticated search and/or multi-start strategies. However, for general SNEs, neither root-finding methods nor global optimization algorithms can guarantee finding one, no less all, solutions.

Meanwhile, in recent years Large Language Models (LLMs) have shown remarkable capabilities in mathematical reasoning and code synthesis [5]. Recent work has also applied LLMs to complex tasks in network science, such as node classification, community detection, and link prediction [19]. This has prompted the current research, an investigation into the effectiveness of LLMs on finding all solutions to new SNEs (SNEs that are neither already documented in the literature nor the internet).

The rest of this paper is organized as follows. In Sect. 2, we formally introduce SNEs and their transformation to global optimization problems in order to find all roots to the SNEs. LLMs are discussed in Sect. 3. In Sect. 4, a suite of four test problems are presented, created using a new approach for SNE generation. In addition, results are showcased on how well nine variants of LLMs are able to find all solutions to the test SNEs. Lastly, Sect. 5 provides concluding remarks and directions for future research.

2 Problem Definition

An SNE can be formally defined as :

$$F_m(x) = \Theta_m \equiv (0,0,\dots,0)^\top \iff \begin{cases} f_1(x_1,x_2,\dots,x_n) = 0 \\ f_2(x_1,x_2,\dots,x_n) = 0 \\ \vdots \\ f_m(x_1,x_2,\dots,x_n) = 0 \end{cases} \tag{1}$$

where $F_m = (f_1, f_2, \dots, f_m) : \mathcal{D}_n \subset \mathbb{R}^n \to \mathbb{R}^m$ consists of continuous real-valued functions with at least one nonlinear component.

Figure 1 provides an graphical representation of the SNEs given by the two transcendental equations in two unknowns, Eqs. (2) – (3), where Eq. (2) is shown in red and Equation (3) is shown in blue. All of the intersection points of the red and blue curves represent the solutions to this SNE.

$$f_1(x_1,x_2) = x_1 - x_1\sin(x_1+5x_2) - x_2\cos(5x_1-x_2) \tag{2}$$

$$f_2(x_1,x_2) = x_2 - x_2\sin(5x_1-3x_2) + x_1\cos(3x_1+5x_2) \tag{3}$$

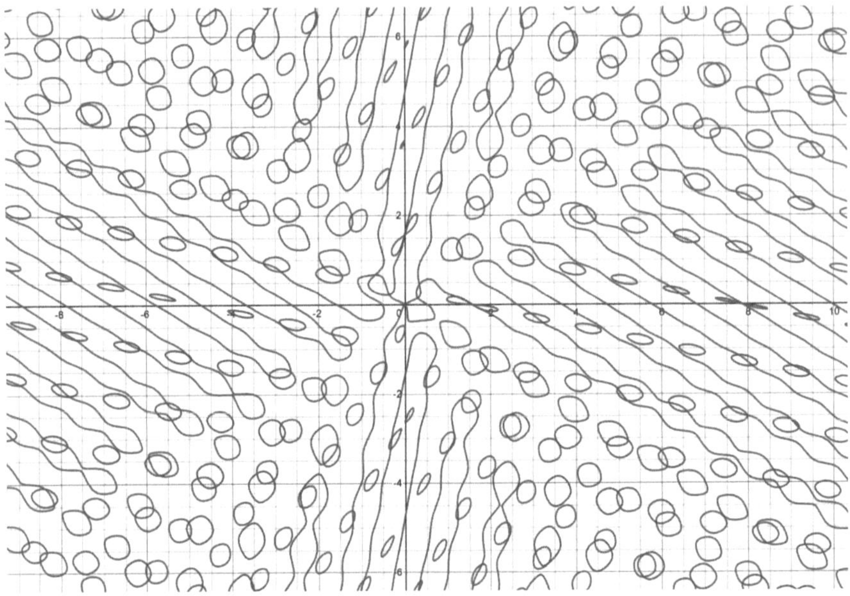

Fig. 1. Example SNE with two equations and two unknowns (Eqs. (2) – (3)). Solutions occur at intersections of the contours.

2.1 Global Optimization Reformulation of SNEs

SNEs can be reformulated as optimization problems to facilitate solution finding. One common reformulation is presented in Equation (4), where $\alpha > 0$, $p > 0$. It is clear that $x^\star$ is a root to the SNE defined by Equation (1) if and only if $\varphi^0(x^\star) = 0$, and $\varphi^0(x) \geq 0$ for all $x \in \mathcal{D}_n$.

$$\arg \underset{x \in \mathcal{D}_n}{min} \, \varphi^0(x) = \alpha \sum_{i=1}^{m} |f_i(x)|^p \tag{4}$$

Assuming k roots have already been found, the generalized penalized objective function for the global optimization reformulation for finding multiple solutions to SNEs is given by Equation (5), where $\varphi^k(x)$ is defined in Equation (6), and $\xi(\delta)$ is continuous and positive for $|\delta| < 1$ and zero elsewhere.

$$\arg \underset{x \in \mathcal{D}_n}{min} \, \varphi(x) = \varphi^0(x) + \varphi^k(x) \tag{5}$$

$$\varphi^k(x) = \alpha \sum_{i=1}^{m} |f_i(x)|^p + \sum_{j=0}^{k} a \, \xi \left(\frac{\|x - x^j\|_2}{\rho} \right) \tag{6}$$

3 Large Language Models for Solving Symbolic–Numeric Problems

Modern LLMs have evolved from purely linguistic tools into general-purpose reasoning engines. They are now routinely deployed in workflows that mix symbolic manipulation with high-precision numerical computation. LLMs are built on parameter-dense transformer architectures [16]. These multi-layer self-attention networks routinely contain tens to hundreds of billions of learned weights and are pre-trained on web-scale corpora that can exceed 10 trillion tokens, spanning books, articles, code, and conversational text. By optimizing the autoregressive objective of next-token prediction, the model gradually internalizes surface-level syntax as well as deeper semantic relationships, factual world knowledge, and heuristic reasoning patterns [18]. The resulting representations can then be specialized through techniques such as supervised fine-tuning or reinforcement learning from human feedback (RLHF), allowing a single backbone model to tackle a wide range of downstream language-understanding and generation tasks with minimal task-specific modifications.

Frontier releases such as LLAMA 3 [15] demonstrate that scaling data diversity is at least as important as scaling parameter count. Although the full training recipe is proprietary, the developers disclose that roughly one-quarter of the training corpus was curated for mathematical reasoning. Despite being trained with a simple next-word objective, LLMs display several additional capabilities that are particularly relevant to mathematical problem solving:

- *Chain-of-thought* induces models to articulate intermediate reasoning steps. This decomposition not only reduces hallucination but enables step-by-step verification, error diagnosis, and alignment with human problem-solving workflows.
- *External tool use* allows the model to emit code snippets (typically Python) that execute code and call external libraries for exact arithmetic, linear algebra, or symbolic algebra, thereby off-loading calculations to a deterministic back-end [14].
- *Self-critique mechanisms* enable models to evaluate their own solutions through reflection prompts (e.g., "Verify this step" or "Identify potential errors"). This provides lightweight uncertainty estimation and can be iteratively applied to refine outputs, mimicking human self-correction processes.

Empirical studies now confirm that such behaviors translate into genuine problem-solving skill. On the MATH benchmark of competition-level questions [5], parameter scaling alone yields a steep accuracy curve. LLMs also perform well on other mathematical benchmarks [12,18]. Similar gains appear on graph-theoretic tasks: a general-purpose LLM can, without gradient updates or additional training, approach specialized graph algorithms on tasks such as node classification and link prediction when fed carefully structured textual descriptions of networks and tasks [17].

4 Testing LLMs for Finding All Roots to SNEs

Accurate benchmarking of SNE solvers demands test instances whose solution sets are known exactly. The *infinite-precision SNE generator* [8] addresses this need by synthesizing equations around user-specified solutions. In addition, by using this SNE generator, there is no chance that any of the LLMs have encountered these problems before, providing a true level-field for performance comparison. For this initial testing, four fairly simple SNEs are generated, enabling a systematic, precision-controlled stress test of LLMbased SNE solvers under a broad spectrum of algebraic and analytic conditions.

4.1 Test Problems

Test Problem 1 (SNE 1). The first test problem SNE is presented in Eqs. (7) – (8). This system contains two equations in two variables, and all equations are polynomial. There is only one solution to this system, given by $(3, -1)$.

$$f_1(x_1, x_2) = -2x_1^2 + 3x_2 + 21 \tag{7}$$

$$f_2(x_1, x_2) = 5x_1x_2^3 - 2x_1^2x_2 + 3x_1x_2 + 6 \tag{8}$$

Test Problem 2 (SNE 2). The second test problem SNE is presented in Eqs. (9) – (10). This system is a modification of the first system by introducing polynomial denominators to each function, resulting in a system of two rational functions in two dimensions. The solution to this system is also $(3, -1)$. However, these two equations are not differentiable across the hypersurfaces $x_1 = x_2$ and $x_1x_2 + 2x_1 - 2x_2 - 4 = 0$, posing a challenge for derivative-based solvers that cannot detect or circumvent such breaks automatically.

$$f_1(x_1, x_2) = \frac{-2x_1^2 + 3x_2 + 21}{x_1 - x_2} \tag{9}$$

$$f_2(x_1, x_2) = \frac{5x_1x_2^3 - 2x_1^2x_2 + 3x_1x_2 + 6}{x_1x_2 + 2x_1 - 2x_2 - 4} \tag{10}$$

Test Problem 3 (SNE 3). The third system, presented in Eqs. (11) – (12) also has the solution $(3, -1)$, but in this system both equations are transcendental. The result is a smooth, but highly nonconvex system.

$$f_1(x_1, x_2) = \sin\left(\tfrac{\pi}{2}x_1\right)\cos\left(\tfrac{\pi}{3}x_2\right) - x_1^2 + x_2 + \tfrac{21}{2} \tag{11}$$

$$f_2(x_1, x_2) = \sin\left(\tfrac{\pi}{6}x_1x_2\right) - x_2\sin\left(\tfrac{\pi}{6}x_1\right) - x_1x_2 - 3 \tag{12}$$

Test Problem 4 (SNE 4). The fourth test problem is a more complex polynomial system (Eqs. (13) – (14)), involving two equations in two unknowns. There are three distinct roots for this system, $(3, -1)$, $(0, 2)$, and $(0, -2)$. This test problem illustrates how the SNE generator [8] can embed multiple isolated solutions in a single system, making it suitable for evaluating a solver's ability to discover *all* roots rather than just one.

$$f_1(x_1, x_2) = -2x_1^4 - 2x_1^2x_2^2 + 3x_1^2x_2 + 29x_1^2 + 3x_2^3 + 21x_2^2 - 12x_2 - 84 \quad (13)$$

$$f_2(x_1, x_2) = -6x_1^4x_2 + 15x_1^3x_2^3 + 9x_1^3x_2 + 8x_1^2x_2^3 - 32x_1^2x_2 + 18x_1^2 - 20x_1x_2^5 + 68x_1x_2^3 + 48x_1x_2 - 24x_2^2 + 96 \quad (14)$$

4.2 LLM-Based Solution Procedure

This section describes how LLMs were employed to solve the four small SNE test problems discussed above.

Prompt Design. Every run is driven by a single, human-readable instruction that names the task unambiguously and lists the equations in algebraic notation. The same template is reused for all four SNE test systems; only the equation block changes.

Prompt Template

```
solve this equation system, print all solutions, nothing else
{equation_1} = 0
{equation_2} = 0
```

Here `{equation_1}` and `{equation_2}` are replaced at runtime by the explicit algebraic expressions of the test SNE.

4.3 Models Evaluated

The study covers nine publicly accessible LLM endpoints, each in the newest revision available at the time of writing:

- **ChatGPT**[1]: `4o` and `o3`.
- **Grok**[2]: "non-thinking" and "thinking" presets.
- **DeepSeek**[3]: "non-thinking" and "thinking" presets.
- **Gemini**[4]: `2.5-flash` and `2.5-pro`.
- **Claude**[5]: latest flagship checkpoint.

---

[1] https://chatgpt.com.
[2] https://grok.com/.
[3] https://www.deepseek.com/.
[4] https://gemini.google.com.
[5] https://claude.ai/.

4.4 Test Results and Discussion

Table 1 shows the results when each LLM is prompted to solve each test SNE instance. For reference, each SNE column has the number of true solutions to that SNE given in parentheses. For each combination of test SNE and LLM, three numbers are listed: the number of correct solutions found (F); the number of correct solutions not found (M); and the number of incorrect solutions found (W). Note that F + M will equal the number of true solutions for the SNE.

Table 1. Results from allowing each LLM tested to solve each test SNE.

	SNE 1 (1)			SNE 2 (1)			SNE 3 (1)			SNE 4 (3)		
LLM	F	M	W	F	M	W	F	M	W	F	M	W
ChatGPT 4o	1	0	0	1	0	0	0	1	0	3	0	4
ChatGPT o3	1	0	2	1	0	2	1	0	2	3	0	4
Grok	0	1	2	0	1	3	0	1	0	0	3	4
Grok (th.)	1	0	0	1	0	0	0	1	0	3	0	0
DeepSeek	1	0	0	1	0	1	1	0	0	2	1	0
DeepSeek-R1	1	0	0	-	-	-	1	0	0	-	-	-
Gemini Flash	1	0	0	1	0	0	1	0	0	2	1	0
Gemini Pro	1	0	6	1	0	0	1	0	0	2	1	9
Claude	0	1	4	0	1	2	1	0	1	0	3	3

F: Number of correct solutions found; **M**: Number of correct solutions not found; **W**: Number of spurious (incorrect) solutions found.

The experiments reveal two distinct solution pathways favored by current LLMs. First, models such as DeepSeek, Gemini, and Grok tend to keep all reasoning *in-context*. They manipulate the SNEs symbolically, aiming for closed-form factors or direct substitution. This analytic route works reliably on the simple SNE, SNE 1, but it struggles whenever the SNEs become more complex. Typical failure modes are skipped variables, sign errors, or an incomplete derivation that terminates when the chain of thought exceeds the context window.

A contrasting behavior emerges from ChatGPT and Claude. These models prefer to create executable Python code that call Python libraries such as `sympy` or `scipy.optimize`. Delegating the heavy computational burden to a deterministic library allows them to locate high-precision roots numerically for each of the tested SNEs, including the highly nonconvex system, SNE 3. The trade-off is an occasional surplus of roots introduced by loose tolerances inside the solver, together with the obvious requirement that code execution be permitted in the environment where the LLM runs.

Across both of the above strategies that the LLMs employed, one glaring weakness is shared: out-of-the-box LLMs seldom verify their own answers. A proposed root from an LLM can fail a direct residual check by the LLM yet still be returned as a solution. Empirically, instructing the LLM to *validate every candidate* almost always removes incorrect roots, but at the cost of a longer dialogue and increased run time.

5 Conclusions and Future Research

LLMs already contain the primitives needed to handle finding roots to SNEs, yet no single architecture or prompting style is universally dominant. Pure analytic reasoning is transparent and lightweight but falters on complex expressions; tool-calling shifts the burden to mature numerical libraries but introduces its own sources of noise. A pragmatic way forward is a hybrid protocol: invoke an LLM with tool access to generate candidate roots, evaluate those roots at arbitrary precision, feed any non-zero residuals back to the model, and iterate until convergence. Future research will focus on tighter symbolicnumeric integration and on training objectives that reward explicit self-verification before an answer is finalized.

References

1. Broyden, C.G.: A class of methods for solving nonlinear simultaneous equations. Math. Comput. **19**(92), 577–593 (1965)
2. Dennis, J.E., Schnabel, R.B.: In: Numerical methods for unconstrained optimization and nonlinear equations. SIAM (1996)
3. Floudas, C.A., Pardalos, P.M.: A collection of test problems for constrained global optimization algorithms. LNCS. Springer (1990)
4. Floudas, C.A., et al.: Handbook of Test Problems in Local and Global Optimization. Springer, US (1999)
5. Hendrycks, D., et al.: Measuring mathematical problem solving with the MATH dataset. In: Proceedings of the 35th Conference on Neural Information Processing Systems (2021)
6. Hilbert, D.: Mathematical problems. Bull. Am. Math. Soc. **8**, 437–479 (1902)
7. Hirsch, M.J., Pardalos, P.M., Resende, M.G.C.: Solving systems of nonlinear equations with continuous GRASP. Nonlinear Anal. Real World Appl. **10**(4), 2000–2006 (2009)
8. Hirsch, M.J., Semenov, A., Trevena, W., Pardalos, P.M.: A method to generate a general system of nonlinear equations with known solutions. submitted (2025)
9. Jansson, C.: An NP-hardness result for nonlinear systems. Reliable Comput. **4**(4), 345–350 (1998)
10. Kotsireas, I., Pardalos, P.M., Semenov, A., Trevena, W., Vrahatis, M.: Survey of methods for solving systems of nonlinear equations, part I: root-finding approaches. arXiv (2022)
11. Kotsireas, I., Pardalos, P.M., Semenov, A., Trevena, W., Vrahatis, M.: Survey of methods for solving systems of nonlinear equations, part II: optimization-based approaches. arXiv (2022)
12. Li, J.: Numinamath: The largest public dataset in ai4maths with 860k pairs of competition math problems and solutions. Hugging Face repository (2024). Dataset available at https://huggingface.co/datasets/numinamath. Accessed 17 July 2025

13. Pei, J., Dražić, Z., Dražić, M., Mladenović, N., Pardalos, P.M.: Continuous variable neighborhood search (c-vns) for solving systems of nonlinear equations. INFORMS J. Comput. **31**(2), 235–250 (2019)
14. Schick, T., et al.: Toolformer: Language models can teach themselves to use tools (2023)
15. Touvron, H., et al.: LLaMA: Open and efficient foundation language models (2023)
16. Vaswani, A., et al.: Attention is all you need (2023)
17. Wang, H., Feng, S., He, T., Tan, Z., Han, X., Tsvetkov, Y.: Can language models solve graph problems in natural language? In: Thirty-seventh Conference on Neural Information Processing Systems (2023)
18. Ye, Y., Huang, Z., Xiao, Y., Chern, E., Xia, S., Liu, P.: Limo: Less is more for reasoning (2025)
19. Zhang, Y., Wang, H., Feng, S., Tan, Z., Han, X., He, T., Tsvetkov, Y.: Can LLM graph reasoning generalize beyond pattern memorization? (2024)

Chisel Verification Assertions

Zhiwei Wan and Nan Zhang(✉)

Institute of Computing Theory and Technology, Xidian University, Xián 710071, China
124031212122@stu.xidian.edu.cn, nanzhang@xidian.edu.cn

Abstract. This study constructs Chisel Verification Assertions (CVA), a new assertion language and formal verification tool. CVA heavily relies on the ChiselTest framework and integrates the powerful expressive power of Propositional Projection Temporal Logic (PPTL). A temporal operator system that adapts to Chisel characteristics is designed independently in CVA, achieving a leapfrog improvement in the expressiveness of the Chisel assertion language. This operator system, following hardware validation logic, can accurately describe complex temporal and functional properties. Several typical cases of Chisel program verification are selected to validate CVA. The results show that CVA can effectively capture logic and temporal errors in design, verifying its technical effectiveness and practicality in hardware verification scenarios.

Keywords: Chisel · Assertion Language · PPTL · Formal Verification

1 Introduction

With the accelerated iteration of integrated circuit technology, the scale of digital circuits is growing rapidly. Chisel, a Scala-based hardware description language, has been widely adopted in digital circuit design. Its concise programming paradigm, combined with flexible abstraction mechanisms, thus allows for deep integration with the software ecosystem. Strong practical engineering capabilities are demonstrated by Chisel in both the development of basic logic modules and the construction of complex System-on-Chip (SoC) systems. As traditional verification methods increasingly struggle to meet the verification requirements, there is a growing need for more efficient and accurate verification methods. The correctness and reliability of Chisel designs must be ensured in complex application scenarios.

Traditional verification frameworks represented by ChiselTest [1] can meet basic functional verification requirements. However, significant limitations are exposed when addressing complex designs. These frameworks rely on Scala's native syntax to build verification logic and lack specialized hardware verification semantic expressions. Consequently, lengthy test codes must be written by developers to describe temporal constraints and functional properties. Verification development costs are thereby increased, and code maintainability is

A. Chinchuluun et al. (Eds.): AAIM 2025, LNCS 16379, pp. 35–46, 2026.
https://doi.org/10.1007/978-981-95-5657-1_4

reduced. Additionally, the inherent incompleteness of the simulation verification mode makes it difficult to achieve ideal verification coverage, and formal verification of the design cannot be performed.

In the development of the Chisel program verification techniques, numerous studies have been conducted to address the limitations of traditional frameworks. In his master's research [2], Tsai extended the syntax of ChiselTest's assertion language and integrated simulation-based verification methods. However, these methods are limited in expressing unbounded sequence repetitions and do not cover formal verification. ChiselVerify [3] adapts the Universal Verification Methodology (UVM) to the Chisel platform, yet its core remains focused on test-driven approaches, leaving formal verification requirements unmet. Although Xiang et al. [4] explored formal verification for Chisel programs, the high coupling between verification specifications and design code was identified as a factor affecting efficiency and maintainability.

CHA [5], proposed by Shizhen Yu et al., was inspired by SystemVerilog Assertions (SVA) to revitalize Chisel verification. This study introduces an innovative upgrade: the Chisel Verification Assertions (CVA) system. By deeply integrating Propositional Projection Temporal Logic (PPTL) [10] into the verification architecture, CVA achieves several key advancements. Leveraging PPTL's expressive temporal operators, CVA not only enables precise modeling of complex temporal relationships but also enhances verification reliability through its rigorous theoretical foundation. Syntactically, user-defined logical expressions are supported by CVA to accommodate diverse verification scenarios. Technologically, CVA's seamless integration with formal methods marks a significant breakthrough in Chisel program verification.

2 Design and Implementation of CVA

2.1 Overall Architecture of CVA

The architecture of CVA consists of three major modules, as shown in Fig. 1, enabling efficient verification of Chisel programs through an innovative design flow. In the input module, developers submit a Chisel design file and a CVA assertion statement. As the core carrier of verification requirements, the timing and functional properties of hardware behaviors can be precisely described by the CVA assertion.

The FIRRTL Compiler module plays a critical role in compilation and transformation. First, Chisel designs are translated into high-level and low-level FIRRTL intermediate representations, which are ultimately constructed into a transition system. During this process, our CVA assertion is encapsulated within an Annotation structure. This encapsulation enables the compiler to automatically establish mappings between FIRRTL constructs and the original Chisel code. The assertion is transformed into a Propositional Projection Temporal Logic (PPTL) formula, providing a formal basis for subsequent verification. Model checking techniques based on automata theory are employed. The generated PPTL formula is converted into a Büchi Automaton (BA) using the pptl2ba tool [7], which represents formula violations in NeverClaim form. Subsequently, the BA is converted into the Higher-order Automata (HOA) standard format using the Spot tool [8] and then adapted into a transition system homologous to the Chisel design. After a product operation, the result is stored in the Btor2 format.

Verification is performed by the bounded model checker BtorMC [9], which examines the system for a specified number of steps. If a non-empty result is detected, BtorMC generates counterexample traces to assist developers in quickly identifying design flaws, thereby achieving in-depth verification of Chisel programs.

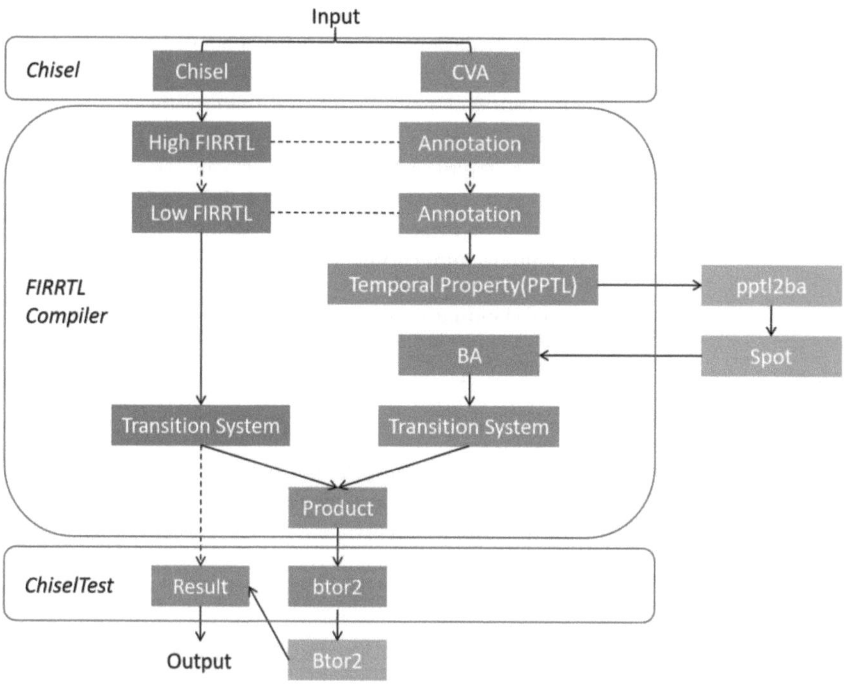

Fig. 1. The overall architecture of CVA.

2.2 Temporal Operator System Design

The temporal operator system of CVA is designed to fully accommodate the Scala-based language features of Chisel and the practical requirements of hardware verification. As shown in Table 1, it encompasses a comprehensive set of operators, including sequence operators and property operators. These operators can precisely describe the temporal relationships of signals, the chronological

Table 1. CVA temporal operators

	Operators	Meaning
property operators	[*n]	**Fixed Repetition:** A sequence repeats consecutively n times.
	[*m:n]	**Range Repetition:** A sequence repeats consecutively between m and n times, inclusive.
	##n	**Fixed Delay:** A sequence that begins after n clock cycles from the current clock cycle.
	##[m,n]	**Range Delay:** A sequence that occurs at some point between m and n clock cycles from the current clock cycle.
	&	**Sequence Intersection:** Both sequences must be valid simultaneously, but are allowed to end asynchronously.
	|	**Sequence Or:** At least one of the two sequences shall hold, allowing them to start or end at different time points.
sequence operators	!	**Property Negation:** The negation of a property.
	X	**Next:** A property holds at the next clock cycle.
	&&	**Logic And:** Both properties are true concurrently.
	||	**Logic Or:** At least one of the two properties must hold.
	U	**Until:**The antecedent holds continuously until the consequent holds.
	G	**Always:** A property holds in all clock cycles.
	F	**Eventually:** A property will hold in some future clock cycle.
	|->	**Overlapping implication:** Verification of subsequent conditions begins in the same clock cycle when the antecedent condition holds.
	|=>	**Non-overlapping implication:** Verification of subsequent conditions begins in the next clock cycle after the antecedent condition holds.

order of events, and complex functional properties in Chisel programs, significantly enhancing the expressive power of the Chisel assertion language.
The sequence expression format is as follows:

$$s := u \mid (s) \mid s\#\#ms \mid s\#\#[m:n]s \mid s|s \mid s\&s \mid s[*m] \mid s[*m:n] \quad (1)$$

where s is a sequence expression and u is a Boolean expression.
The property expression format is as follows :

$$p := s \mid (p) \mid s|- > p \mid s| => p \mid !p \mid Gp \mid Fp \mid Xp \mid pUp \mid p||p \mid p\&\&p \quad (2)$$

where p is a property expression and s is a sequence expression.

2.3 Integration of PPTL

As an extension of Propositional Interval Temporal Logic (PITL) [11], Propositional Projection Temporal Logic (PPTL) achieves significant breakthroughs in the original logical system. PPTL extends the applicability of PITL from finite to infinite intervals, remarkably enhancing the completeness of logical expression. In terms of expressive power, PPTL reaches full regularity [6], making it more adept at describing complex temporal properties, particularly interval-related characteristics and periodic repetition properties.

The deep integration of PPTL with the Chisel assertion language is realized by defining the mapping relationships between PPTL and Chisel data types and operations, enabling the effective application of PPTL in the Chisel verification environment. For example, propositions in PPTL correspond to signal states in Chisel, and the temporal logic of PPTL is utilized to describe the changes of signal states over time.

Syntax. The PPTL formula P is as follows :

$$P ::= p \mid \bigcirc P \mid \neg P \mid P_1 \vee P_2 \mid (P_1, \ldots, P_m) prj\ P \tag{3}$$

Let *Prop* be a non-empty set of atomic propositions, where $p \in Prop$. $P_1, \ldots, P_m$ and P are PPTL formulas. The temporal operators include $\bigcirc$ (next) and prj (projection).Operators $\neg$, $\vee$ in classical logic are also applicable in PPTL.

Semantics. We explain the semantics of PPTL from three parts: states, intervals and interpretations.

States. A mapping from $Prop$ to $B = true, false$ is defined as a state s on Prop:

$$s : Prop \rightarrow B \tag{4}$$

$s[p]$ denotes the value of atomic proposition p on state s.

Intervals. The interval σ is defined as a sequence of states, which is not empty and can be finite or infinite. $|\sigma|$ denotes the length of a interval. The length of the finite interval is the number of states contained in the interval minus one. The length of the infinite interval is denoted by ω. We define ω=ω and $i < \omega$ for any element i in the set N_0 of nonnegative integers. To simplify the definition, $< s_0, \ldots, s_{|\sigma|} >$ is used to represent a continuous state interval of length $|\sigma|$.

Interpretations. The PPTL formula is interpreted as a triple $I = (\sigma, k, j)$, where σ is an interval, k is an integer and j is an integer or ω. σ, k and j satisfy the relation $k \preceq j \leq |\sigma|$. $\preceq$ is defined as $\leq -\{(\omega, \omega)\}$. The symbol $(\sigma, k, j)| = P$ represents that the formula P is satisfied on the subinterval $< s_k, \ldots, s_j >$ of σ, where the state of the interval is s_k.

Expansion of Connectives. PPTL has the following derived formulas:

$$\begin{array}{ll}
\varepsilon \stackrel{\text{def}}{=} \neg \bigcirc true & \bar{\varepsilon} \stackrel{\text{def}}{=} \neg \varepsilon \\
\bigcirc^0 P \stackrel{\text{def}}{=} P & \bigcirc^n P \stackrel{\text{def}}{=} \bigcirc(\bigcirc^{n-1} P) \\
len(n) \stackrel{\text{def}}{=} \bigcirc^n \varepsilon & skip \stackrel{\text{def}}{=} len(1) \\
\odot \stackrel{\text{def}}{=} \varepsilon \vee \bigcirc P & P;Q \stackrel{\text{def}}{=} (P;Q) prj\ \varepsilon \\
\Diamond P \stackrel{\text{def}}{=} true; P & \Box P \stackrel{\text{def}}{=} \neg \Diamond \neg P
\end{array} \tag{5}$$

In the above operators, $\odot$ (weak next), $\Box$ (always), $\Diamond$ (sometimes) and ; (chop) are the derived timing operators. ε (empty) denotes the interval of length 0. $\bar{\varepsilon}$ (more) indicates that the current state of the interval is not the final state. Operators $true$, $false$, $\wedge$, $\rightarrow$ and $\leftrightarrow$ in classical logic are also applicable in PPTL.

Table 2. Precedence rules

Precedence	Operators
1	$\neg$
2	$\bigcirc$, $\odot$, $\Diamond$, $\Box$
3	$\wedge$, $\vee$
4	$\rightarrow$, $\leftrightarrow$
5	prj, ;

Precedence Rules. The precedence rules are used as shown in Table 2, where 5 = lowest and 1 = highest.

Satisfaction and Validity. For a certain interval σ, if $(\sigma, 0, \sigma)| = P$ holds, then the formula P is satisfiable in this interval. If P is satisfied in any interval, then P is valid.

The Mapping Relationship Between CVA Operators and PPTL Operators. The mapping relationship between CVA operators and PPTL operators is shown in Table 3.

In the table, s, s_1, and s_2 denote sequences in CVA assertions. p, p_1, and p_2 represent properties in CVA assertions. P, P_1, and P_2 are PPTL formulas corresponding to these sequences or properties in CVA assertions. Notably, when translating a CVA assertion into a PPTL formula based on the table's mapping rules, the resulting PPTL formula requires adjustment. The goal is to get a new, logically-equivalent formula. The optimized formula is then fed into pptl2ba to generate a BA. This method simplifies the formula structure, making it compatible with the automaton conversion algorithm and easing subsequent processing. For instance, moving the negation operator ! as close to atomic propositions flattens the formula, reduces nesting levels, and simplifies the analysis of atomic proposition state transitions. Such simplification cuts down on computational resources and time, boosting the translation process's efficiency and accuracy.

2.4 Passing Assertions to FIRRTL and Preprocessing

Chisel and FIRRTL are complementary hardware design tools. Chisel empowers developers with high-level abstract programming capabilities. FIRRTL, as a core intermediate representation, not only undertakes the critical task of converting Chisel code into hardware implementations but also possesses optimization

Table 3. The mapping relationship between CVA operators and PPTL operators

CVA operators	PPTL operators
$s\ [*n]$	$P \wedge len\ n \wedge \Box(\bigcirc more \rightarrow \bigcirc P)$
$s\ [*m : n]$	$P \wedge len\ m \wedge \Box(\bigcirc more \rightarrow \bigcirc P)$
$s_1\ \#\#n\ s_2$	$P_1 \rightarrow \bigcirc^n P_2$
$s_1\ \#\#[m : n]\ s_2$	$P_1 \rightarrow \bigvee_{i=m}^{n} P_2$
$s_1\ \&\ s_2$	$P_1\ and\ P_2$
$s_1 \mid s_2$	$P_1\ or\ P_2$
$!\ p$	$!\ P$
$X\ p$	$\bigcirc\ P$
$p_1\ \&\&\ p_2$	$P_1\ and\ P_2$
$p_1 \mid\mid p_2$	$P_1\ or\ P_2$
$p_1\ U\ p_2$	$P_1\ U\ P_2$
$G\ p$	$\Box\ P$
$F\ p$	$\Diamond P$
$p_1 \mid -> p_2$	$P_1 \rightarrow \Diamond P_2$
$p_1 \mid => p_2$	$P_1 \rightarrow \bigcirc \Diamond P_2$

and extension functions. The organic combination of the two establishes an efficient and comprehensive hardware design system. In CVA, Chisel assertions are encapsulated within FIRRTL's annotation structures. The FIRRTL compiler maintains a consistent mapping between signal and construct names throughout the entire code transformation process, as also noted in [5].

2.5 Generating Transition Systems from Büchi Automata

In the verification process, the processed PPTL formulas are fed into the external tool pptl2ba [7], which outputs a BA represented in NeverClaim format. Subsequently, this BA is converted via the Spot tool [8] into a deterministic BA in HOA format. Finally, BA is transformed into a Transition system, which has the same data structure as the Transition system designed by Chisel.

2.6 Transition System Product

A Transition system is composed of the following key components:

(1) **name**: Identification of the system.
(2) **inputs**: A list of BVSymbol elements representing input signals, where BVSymbol denotes a bit-vector symbol.
(3) **states**: A collection of states presented as a list of State types.
(4) **signals**: A list of signals consisting of Signal elements.

(5) **comments (optional)**: A Map used to store annotation information related to the Transition system.
(6) **header (optional)**: A string with a default empty value, often serving as the header information for the Transition system.

When performing the product operation on two Transition systems, the Transition system derived from the Chisel design serves as the basis. The states and signal sets from the Transition system generated by the CVA assertion are then incorporated to obtain the product of these two Transition systems.

2.7 Property Verification

The Transition system product is stored in the Btor2 file format to facilitate property verification by BtorMC [9]. Counterexamples indicating assertion violations can be visualized in the form of waveform diagrams.

3 Case Studies

To evaluate the effectiveness and practicality of CVA, two representative Chisel program verification cases were selected: a counter module and a greatest common divisor (GCD) solver module.

3.1 4-Bit Counter Module

This module functions as a 4-bit counter that starts from 0, increments until reaching 15, and then remains unchanged. We aim to verify whether the counter, starting from 1, can maintain the incrementing state for exactly 15 cycles. This property can be formalized as the following CVA assertion:

$$counter1 \ |-> \ isIncrementalCounting\ [*15] \tag{6}$$

The signal $counter1$ is asserted when $counter === 1.U$, indicating the count value is 1. Here, the $counter$ is the output signal representing the current count. The incremental counting is indicated by the signal $isIncrementalCounting$. The signal is asserted when $counter - 1.U === RegNext(counter)$, meaning that the count value minus 1 in the current cycle equals the count value in the previous cycle. A successful test confirms that the incrementing state remains stable throughout this process. To verify if the counter can maintain the incrementing state for 16 cycles starting from 1, the property can be formalized as:

$$counter1 \ |-> \ isIncrementalCounting\ [*16] \tag{7}$$

In this case, the test fails and a counterexample is obtained, as shown in Fig. 2. After initialization is completed in the first cycle, counting begins. When the count reaches 1, both $counter1$ and $isIncrementalCounting$ are asserted. Subsequently, the high level of $isIncrementalCounting$ is maintained for 15 cycles,

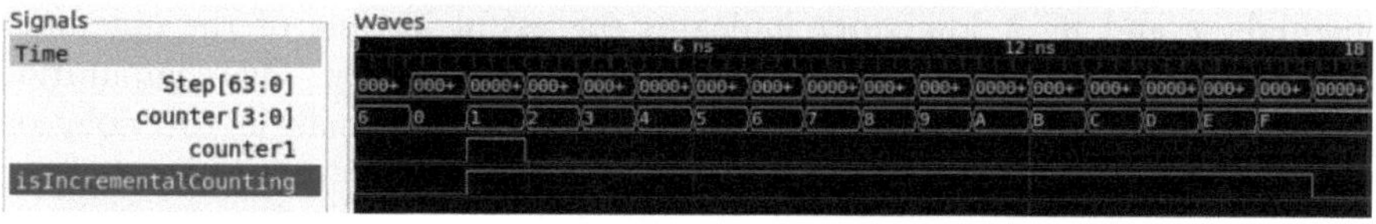

Fig. 2. The counterexample for $counter1\ |-> \ isIncrementalCounting\ [*16]$

and the count increments to 15 before remaining unchanged. At this point, $isIncrementalCounting$ becomes low, demonstrating that the property is violated.

To verify the stability of the counter's incrementing behavior, the following assertion can be used to check that the incrementing state remains stable throughout all 16 counting cycles:

$$isIncrementalCounting\ |-> \ isIncrementalCounting\ U\ counter16 \tag{8}$$

The signal $counter16$ is asserted in the second cycle after $counter\ === \ 15.U$ holds. If we modify this module to a 4-bit binary up-counter, it can achieve cyclic counting from 0 to 15. The following CVA assertion verifies that the incrementing functionality holds true indefinitely within the specified number of checked steps, ensuring the stability of the counter:

$$isIncrementalCounting\ |-> \ G\ isIncrementalCounting \tag{9}$$

3.2 4-Bit Input Greatest Common Divisor (GCD) Module

This module computes the GCD of two 4-bit input operands, x and y. Its underlying principle involves repeatedly subtracting the smaller number from the larger one and updating the larger operand with the result, continuing this process until one of the operands becomes zero. At that point, the remaining non-zero operand is the GCD. Our objective is to use CVA assertions to measure the number of cycles required to complete this computation. First, we define an output signal $start$ and a signal end defined as $!start$. When the $start$ is low, the two 4-bit input operands x and y are loaded. During the GCD computation, the $start$ remains high, and the transitions are low once the computation is complete. To verify whether computing the GCD for any two 4-bit inputs x and y requires at most 10 cycles, we can formalize this property as the following CVA assertion:

$$start\ |-> \ \#\#[1:10]\ end \tag{10}$$

This assertion specifically means that the end must hold within 1 to 10 cycles after the $start$ is asserted. The test fails, and a counterexample is generated, as shown in Fig. 3. The first two cycles are for initialization and loading the input operands x and y. A low *start* indicates successful loading of the operands. In the counterexample, x and y are 1 and 10, respectively. The computation completes when x becomes 0, requiring 11 cycles, which confirms the violation of the property.

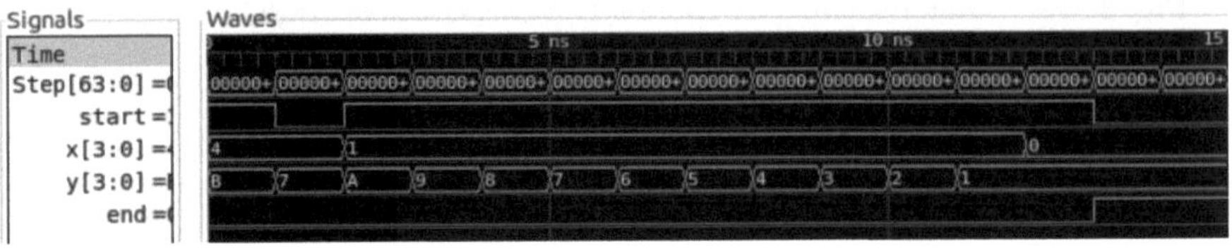

Fig. 3. The counterexample for $start\ |-> \ \#\#[1:10]\ end$.

If we want to verify that computing the GCD of any two 4-bit input operands x and y requires at most 15 cycles, this property can be formalized as the following CVA assertion:

$$start\ |-> \ \#\#[1:15]\ end \tag{11}$$

The test fails, and a counterexample is generated as shown in Fig. 4. The computation requires 16 cycles in this case, confirming the violation of the property. This result also demonstrates that CVA can generate different counterexamples based on varying assertions, aiding in error localization for Chisel program designs. Note that the assertion checker only reports counterexamples within the specified number of steps and does not generate longer ones.

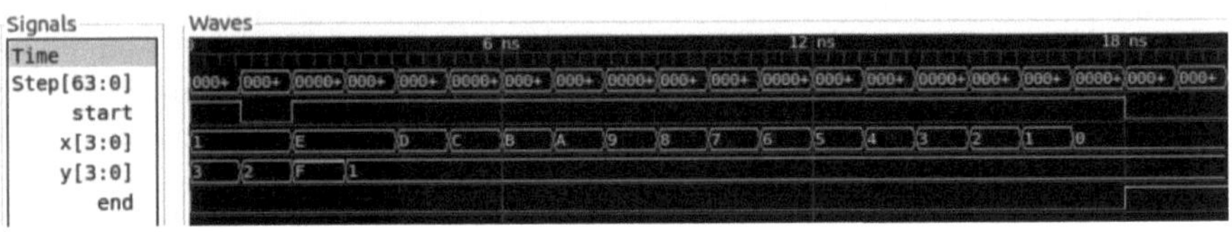

Fig. 4. The counterexample for $start\ |-> \ \#\#[1:15]\ end$.

In fact, computing the GCD of two 4-bit input operands x and y requires at most 16 cycles. Therefore, the corresponding CVA assertion should be:

$$start\ |-> \ \#\#[1:16]\ end \tag{12}$$

Then the test will pass without any counterexamples. Additionally, we can use the following assertion to verify that once the computation starts, the GCD of x and y will eventually be computed:

$$start\ |-> \ F\ end \tag{13}$$

Table 4. Performance Comparison of CVA with Existing Tools Across Different Cases

Verification Case	Verification Tool	Runtime (s)	Memory Usage (MB)	Verification Depth (State Steps)	Supported Temporal Property Types
4-bit Counter	CVA	1.2	45	128	Infinite Loop, Fixed Delay
	ChiselTest	0.8	32	64	Finite Steps Only (No Infinite Properties)
	CHA	1.5	51	128	Fixed Delay, Sequence Repetition
4-bit GCD Module	CVA	2.3	68	256	Range Delay, Eventual Property
	ChiselTest	1.9	55	128	Finite Steps Only (No Infinite Properties)
	CHA	2.7	72	256	Range Delay, Sequence Repetition

3.3 Benchmark Test for Verification Tool Performance Comparison

To further quantify the performance of CVA in practical applications, we conducted a benchmark test for performance comparison. Three core metrics, namely runtime, memory usage, and verification depth (state steps), were selected to compare CVA with ChiselTest and CHA. The results are presented in Table 4.

This result indicates that CVA maintains competitiveness in performance compared with existing tools, without significant performance overhead caused by the integration of PPTL. Both CVA and CHA achieve a verification depth of 128 state steps (for the 4-bit counter) and 256 state steps (for the 4-bit GCD module), which is significantly higher than ChiselTest's 64 and 128 state steps, respectively.

The key difference lies in the type of property support: Limited by its "simulation-based verification mode," ChiselTest can only handle functional properties with finite steps and cannot describe complex temporal logics such as "infinite loops" and "eventual properties." Although CHA supports partial temporal properties, it cannot fully capture the regular expressive capability of PPTL. CVA achieves a balance among "performance, expressiveness, and flexibility," making it applicable not only to small and medium-sized cases but also laying a performance foundation for the verification of complex designs in subsequent work.

4 Conclusion and Future Work

This research successfully developed CVA. The expressive power of Chisel's assertion language is significantly enhanced by deeply integrating the ChiselTest framework, PPTL, and a self-designed temporal operator system. Case studies demonstrate that CVA is technically feasible and practical in hardware verification, facilitating more accurate system designs.

Despite these achievements, there remains room for further optimization and expansion. Future research will focus on several key directions. First, verification efficiency will be improved by refining CVA's underlying algorithms to handle larger-scale and more complex Chisel designs. Second, integration with other advanced verification techniques, such as hybrid approaches combining formal model checking and simulation, will be explored to enhance verification coverage. Finally, practical applications will be strengthened by integrating CVA with more open-source Chisel projects and industrial designs, continuously refining its capabilities and performance.

Acknowledgments. This research was supported by the National Natural Science Foundation of China under Grant Nos. 62272359, 62192734 and 62172322; Natural Science Basic Research Program of Shaanxi Province under Grant Nos. 2023JC-XJ-13 and 2022JM-367.

References

1. Laeufer, K., Bachrach, J., Sen, K.: Open-source formal verification for Chisel. In: WOSET (2021)
2. Tsai, Y.C.A.: Dynamic verification library for chisel. Master's thesis, University of California, Berkeley (2021)
3. Dobis, A., et al.: ChiselVerify: an open-source hardware verification library for chisel and scala. In: NorCAS, pp. 1–7. IEEE (2021)
4. Xiang, M., Li, Y., Tan, S., Zhao, Y., Chi, Y.: Parameterized design and formal verification of multi-ported memory. In: ICECCS, pp. 33–41. IEEE (2022)
5. Yu, S., et al.: CHA: Supporting SVA-Like Assertions in Formal Verification of Chisel Programs (Tool Paper). Springer, Cham (2022)
6. Duan, Z., Tian, C., Zhang, L.: A decision procedure for propositional projection temporal logic with infinite models. Acta Informatica **45**(1), 43–78 (2008)
7. Zhang, X.M.: Implementation and Verification Examples of Partial Order Model Checker for PPTL. Master's Thesis, Xidian University (2015)
8. Duret-Lutz, A., Lewkowicz, A., Fauchille, A., Michaud, T., Renault, É., Xu, L.: Spot 2.0—a framework for LTL and ω-automata manipulation. In: Artho, C., Legay, A., Peled, D. (eds.) ATVA 2016. LNCS, vol. 9938, pp. 122–129. Springer, Cham (2016)
9. Niemetz, A., Preiner, M., Wolf, C., Biere, A.: Btor2, BtorMC and Boolector 3.0. In: Chockler, H., Weissenbacher, G. (eds.) CAV 2018. LNCS, vol. 10981, pp. 587–595. Springer, Cham (2018)
10. Duan, Z., Koutny, M., Holt, C.: Projection in temporal logic programming. In: Pfenning, F. (ed.) LPAR 1994. LNCS, vol. 822, pp. 333–344. Springer, Heidelberg (1994). https://doi.org/10.1007/3-540-58216-9_48
11. Moszkowski, B.: Compositional reasoning about projected and infinite time. In: Proceedings of First IEEE International Conference on Engineering of Complex Computer Systems (ICECCS '95), pp. 238–245. IEEE (1995)

Approximation Algorithms for Monotone k-Submodular Maximization Under the Chance Constraint

Qiao Chen[1], Shufang Gong[2], and Bin Liu[1](✉)

[1] School of Mathematical Sciences, Ocean University of China, Qingdao 266100, China
chenqiao@stu.ouc.edu.cn, binliu@ouc.edu.cn

[2] School of Mathematics and Information Science, Shandong Technology and Business University, Yantai 264005, China
202514467@sdtbu.edu.cn

Abstract. Submodular optimization is an important topic in the field of combinatorial optimization. The k-submodular function is a generalization of the submodular function. Many machine learning problems, such as placing k types of sensors, can be formulated as maximizing a monotone k-submodular function. In real-world optimization problems where the constraints involve random components, chance constraints are often used to limit the probability of constraint violations due to random factors. In this paper, we investigate the problem of maximizing a monotone k-submodular function under the chance constraint. Specifically, each element e is assigned a random weight independently sampled from a uniform distribution. We study two distinct cases, in which the weights are drawn from a common uniform distribution or from element-specific uniform distributions, all with the same dispersion. For the first case, this paper designs a threshold-decreasing algorithm with an approximation ratio of $1/2-\epsilon$. For the second case, this paper first designs a deterministic algorithm, which uses the method of dividing the ground set into two parts, and then combines its output solution with the threshold-greedy algorithm to design an algorithm with an approximation ratio of $1/3-\epsilon$.

Keywords: k-Submodular maximization · Chance constraints · Decreasing-threshold procedure · Threshold-greedy algorithm

1 Introduction

Submodular optimization has become a research hotspot in recent years due to the property of *submodular functions* as set functions with diminishing returns. Submodular functions naturally arise in numerous classical combinatorial optimization problems. For instance, the rank function of matroids [2] and the cut function of graphs [17] are both submodular. Furthermore, submodularity is crucial in various real-world applications, including facility location [3] and document summarization [11]. As a result, the submodular maximization problem has

A. Chinchuluun et al. (Eds.): AAIM 2025, LNCS 16379, pp. 47–58, 2026.
https://doi.org/10.1007/978-981-95-5657-1_5

been widely studied. Common constraints include cardinality constraints [1,12], knapsack constraints [9,18], and matroid constraints [2,10], etc. Under the cardinality constraint, Nemhauser et al. [12] proposed the classic greedy algorithm in 1978 and proved that it guarantees an approximation ratio of $1 - 1/e$. Later, Feige [6] further proved in 1998 that under the premise of P≠NP, the above approximation ratio is tight. Under the knapsack constraint, Sviridenko [18] in 2004 proposed an algorithm that combines a partial enumeration with a greedy strategy, and proved that it achieves the tight approximation ratio of $1 - 1/e$. For the matroid constraint, Calinescu et al. [2] combined continuous greedy and pipage rounding technology to propose a randomized algorithm that achieves the optimal approximation ratio of $1-1/e$ in 2011. However, in many real-world optimization problems, constraints often involve uncertainties, and in safety-critical or high-cost production systems, it is essential to limit the probability of constraint violations arising from such stochastic factors. Constraints that explicitly take into account the probability of violation are called *chance constraints.* Based on this concept, the *chance-constrained submodular maximization problem* aims to maximize a given submodular function while requiring that the probability of constraint violation does not exceed a predefined threshold. Consequently, this problem has been extensively studied.

While submodular functions have been widely applied across various domains, many complex real-world scenarios involve objectives that cannot be captured by a single set-based submodular function. Typical examples include the placement of multiple types of sensors, multi-topic influence maximization, and multi-class feature selection. These problems often require modeling multiple disjoint types of elements, motivating the need to generalize the traditional definition of submodular functions from sets to k-sets. Therefore, we are motivated to study the k*-submodular* maximization problem as an extension of the classic submodular maximization problem. In this generalization, the input consists of k mutually disjoint subsets of the ground set rather than a single subset. Since most k-submodular maximization problems are NP-hard, they are generally intractable in polynomial time. Common constraints in these problems include cardinality constraints [14–16], knapsack constraints[8,19–21] and others.

Although recent studies have investigated chance-constrained submodular maximization, to the best of our knowledge, the case of k-submodular functions has not been studied. In this paper, we extend the problem to *chance-constrained* k*-submodular maximization.* This problem can be viewed as a generalization of the deterministic knapsack setting, where the total weight of any feasible solution must not exceed the knapsack capacity. By comparison, in our setting each element e has a random weight $W(e)$ independently drawn from a uniform distribution, either from a common interval $[a - \delta, a + \delta]$ or from element-specific intervals $[a_e - \delta, a_e + \delta]$, all with the same dispersion, where a_e and δ are constants. The goal is to maximize a monotone k-submodular function while ensuring the probability of violating the knapsack constraint does not exceed a given threshold. A main difficulty arises because, when random element weights are independently drawn from uniform distributions, computing the exact distribution of the total weight via convolution quickly becomes impractical as the number of elements increases. To overcome this, we employ surrogate functions for the chance constraint and design threshold-based algorithms accordingly.

1.1 Related Work

In 2019, Chen and Maehara [4] first studied the chance-constrained submodular knapsack problem by generalizing the objective from a linear function to a monotone submodular function. For the case where element weights follow a normal distribution, they transform the chance constraint into a multidimensional knapsack constraint, and show that the expected profit of the obtained solution reaches at least $1 - 1/e - O(\epsilon)$ of the optimal solution. In 2020, Doerr et al. [5] analyzed random element weights from uniform distributions $[a - \delta, a + \delta]$ and $[a_e - \delta, a_e + \delta]$, and showed that the greedy algorithm remains effective under chance constraints by using surrogate functions based on the Chernoff bound and Chebyshev's inequality, achieving approximation ratios of $(1-o(1))(1-1/e)$ and $(1/2 - o(1))(1 - 1/e)$, respectively. In 2022, Gong et al. [7] considered the same problem under the streaming model and designed single-pass streaming algorithms that achieve approximation ratios of $1/2 - \epsilon$ and $1/3 - \epsilon$, respectively. In 2023, Yan et al. [22] first considered the case where the random weights of elements are sampled from a uniform distribution with the same expectation but different dispersion $[a-\delta_e, a+\delta_e]$, and they analyzed the generalized greedy algorithm (GGA) and generalized greedy+Max algorithm (GGMA), proving approximation ratios of $(1/2-o(1))(1-1/e)$ and $1/2-o(1)$, respectively. We summarize these results in Table 1 for clarity. In addition, multi-objective evolutionary algorithms have also been analyzed and applied to submodular problems with chance constraints [13,23].

Table 1. Approximation algorithms for chance-constrained submodular maximization.

Reference	Distribution	Setting	Approximation
COCOON'19 [4]	Normal	Offline	$1 - 1/e - O(\varepsilon)$
AAAI'20 [5]	Uniform $[a - \delta, a + \delta]$	Offline	$(1 - o(1))(1 - 1/e)$
AAAI'20 [5]	Uniform $[a_e - \delta, a_e + \delta]$	Offline	$(1/2 - o(1))(1 - 1/e)$
FAW'22 [7]	Uniform $[a - \delta, a + \delta]$	Streaming	$1/2 - \varepsilon$
FAW'22 [7]	Uniform $[a_e - \delta, a_e + \delta]$	Streaming	$1/3 - \varepsilon$
ECAI'23 [22]	Uniform $[a - \delta_e, a + \delta_e]$	Offline (GGA)	$(1/2 - o(1))(1 - 1/e)$
ECAI'23 [22]	Uniform $[a - \delta_e, a + \delta_e]$	Offline (GGMA)	$1/2 - o(1)$

1.2 Our Contribution

We develop efficient algorithms for chance-constrained k-submodular maximization under two uniform weight settings: $[a - \delta, a + \delta]$ and $[a_e - \delta, a_e + \delta]$. To reduce the query complexity of our main algorithm, we first estimate the range of the optimal value via an auxiliary algorithm, and then use this to determine its threshold. Specifically, in the first case $[a-\delta, a+\delta]$, we propose Algorithm 1, which is only used to estimate the range of the value of the optimal solution. Then, based on the result of Algorithm 1, we design a threshold-decreasing algorithm for our research problem, namely Algorithm 3, which can achieve a $(1/2-\epsilon)$-approximation and has a query complexity of $O(\frac{nk}{\epsilon}\log\frac{1}{\epsilon})$. For the second case $[a_e-\delta, a_e+\delta]$, we design two algorithms for our research problem. First, we design Algorithm 3 to seek the output solution by partitioning the ground set into two parts, which can achieve a 1/11-approximation, and its query complexity is $O(nk)$. Then we combine the output solution of Algorithm 3, the thresholding strategy and post-processing to design Algorithm 4, which can achieve a $(1/3-\epsilon)$-approximation and the query complexity is $O(\frac{nk}{\epsilon}\log\frac{1}{\epsilon})$.

Organization. Section 2 reviews the fundamental concepts of k-submodular functions, introduces the problem under study, and presents the surrogate function for the chance constraint. Sections 3 and 4 describe the design and performance guarantees of our algorithms. Finally, some concluding remarks are given in Sect. 5.

2 Preliminaries

We provide the necessary preliminaries on k-submodular functions, describe the problem formulation, and present some basic results needed for our analysis.

Let $V=\{e_1,e_2,\ldots,e_n\}$ be a ground set and k be an integer. We define $[k]=\{1,2,\ldots,k\}$. Let $(k+1)^V=\{(V_1,V_2,\ldots,V_k)\mid V_i\subseteq V\ \forall i\in[k], V_i\cap V_j=\emptyset\ \forall i\neq j\}$ be a family of k disjoint subsets of V, called a *k-set*, and let $\mathbf{0}=(\emptyset,\ldots,\emptyset)$ denote the empty k-set. For $\boldsymbol{x}=(X_1,X_2,\ldots,X_k)\in(k+1)^V$, we define *support set* $\mathrm{supp}(\boldsymbol{x})=\cup_{i\in[k]}X_i$, X_i as ith set of $\boldsymbol{x}$. Denote by $\boldsymbol{x}(e)$ the position of e in $\boldsymbol{x}$, if $e\in X_i$, then $\boldsymbol{x}(e)=i$, otherwise $\boldsymbol{x}(e)=0$. Adding an element $e\notin\mathrm{supp}(\boldsymbol{x})$ to X_i is denoted by $\boldsymbol{x}\sqcup(e,i)$. We also write $\boldsymbol{x}=\{(e_1,i_1),(e_2,i_2),\ldots,(e_t,i_t)\}$ for $e_j\in\mathrm{supp}(\boldsymbol{x}), i_j=\boldsymbol{x}(e_j), \forall 1\leq j\leq t$. For $\boldsymbol{x}=(X_1,X_2,\ldots,X_k)$ and $\boldsymbol{y}=(Y_1,Y_2,\ldots,Y_k)$, we denote by $\boldsymbol{x}\preceq\boldsymbol{y}$ iff $X_i\subseteq Y_i, \forall i\in[k]$, and let $\boldsymbol{y}\backslash\boldsymbol{x}=(Y_1\backslash X_1, Y_2\backslash X_2,\ldots,Y_k\backslash X_k)$.

Definition 1. [8] *The function $f:(k+1)^V\mapsto\mathbb{R}_+$ is k-submodular iff for any $\boldsymbol{x}=(X_1,X_2,\ldots,X_k)$ and $\boldsymbol{y}=(Y_1,Y_2,\ldots,Y_k)\in(k+1)^V$, we have:*

$$f(\boldsymbol{x})+f(\boldsymbol{y})\geq f(\boldsymbol{x}\sqcap\boldsymbol{y})+f(\boldsymbol{x}\sqcup\boldsymbol{y}),$$

where

$$\boldsymbol{x}\sqcap\boldsymbol{y}=(X_1\cap Y_1,\ldots,X_k\cap Y_k),$$

$$\boldsymbol{x}\sqcup\boldsymbol{y}=\left(X_1\cup Y_1\backslash\left(\bigcup_{j\neq 1}X_j\cup Y_j\right),\ldots,X_k\cup Y_k\backslash\left(\bigcup_{j\neq k}X_j\cup Y_j\right)\right).$$

We denote by $\Delta_{e,i}f(\boldsymbol{x}) = f(X_1, ..., X_i \cup \{e\}, \dots, X_k) - f(X_1, ..., X_i, \dots, X_k)$ the *marginal gain* of adding element e to the set X_i with respect to f. The k-submodularity of f implies the *orthant submodularity*, i.e., $\Delta_{e,i}f(\boldsymbol{x}) \geq \Delta_{e,i}f(\boldsymbol{y})$ for any $\boldsymbol{x}, \boldsymbol{y} \in (k+1)^V$ with $\boldsymbol{x} \preceq \boldsymbol{y}, e \notin \bigcup_{\ell \in [k]} Y_\ell$ and $i \in [k]$. In this paper, we consider f is monotone, i.e., $\Delta_{e,i}f(\boldsymbol{x}) \geq 0$. We also assume that f is normalized, i.e., $f(\mathbf{0}) = 0$.

Our problem can be reformulated as:

$$\max_{\boldsymbol{x} \in (k+1)^V} \quad f(\boldsymbol{x})$$

$$\text{s.t.} \qquad \Pr[W(\mathrm{supp}(\boldsymbol{x})) > B] \leq \alpha.$$

where $f : (k+1)^V \mapsto \mathbb{R}_+$ is a normalized monotone k-submodular function, and α is the upper bound of the probability of violating the constraint. In this paper, we set $\alpha \leq 0.5$. If the probability of violating the constraint is too large, the constraint loses its practical meaning, and the quality of the solution cannot be guaranteed. B is a deterministic constraint boundary. The weights of the elements in the support set are random, and the random weights of the elements are sampled from the uniform distribution $[a - \delta, a + \delta]$ or $[a_e - \delta, a_e + \delta]$, and $W(\mathrm{supp}(\boldsymbol{x})) = \sum_{e \in \mathrm{supp}(\boldsymbol{x})} W(e)$ is a linear function.

Theorem 1. [5] *(Multiplicative Chernoff bound). Assume that $X_1, \dots, X_n$ are independent random variables bounded in the interval $[0, 1]$, and let $X = \sum_{i=1}^{n} X_i$. For any $\epsilon \in [0, 1]$, it holds that*

$$\Pr[X \geq (1+\epsilon)E[X]] \leq \exp\left(-\frac{\epsilon^2 E[X]}{3}\right).$$

The following surrogate function is constructed using the Chernoff bound to satisfy the chance constraint.

Corollary 1. [5] *Let each $W(e)$ with $e \in V$ be chosen independently and uniformly within $[a_e - \delta, a_e + \delta]$. For every $X \subseteq V$, if*

$$B - E[W(X)] \geq \sqrt{3\delta|X| \ln(1/\alpha)},$$

then $\Pr[W(X) > B] \leq \alpha$.

Lemma 1. *$X_1, X_2, ..., X_n$ are n mutually independent random variables, and X_i obeys a uniform distribution on the interval $[a_i - \delta_i, a_i + \delta_i]$, that is, $X_i \sim U(a_i - \delta_i, a_i + \delta_i), i = 1, ..., n$, let $S_n = \sum_{i=1}^{n} X_i$, then the probability distribution of S_n is symmetric about $\sum_{i=1}^{n} a_i$.*

By combining Lemma 1 with the chance constraint, we derive the following result.

Lemma 2. *If for any k-set $\boldsymbol{x}$, $\Pr[W(\mathrm{supp}(\boldsymbol{x})) > B] \leq \alpha$ holds, where $\alpha \leq 0.5$, and for any $e \in \mathrm{supp}(\boldsymbol{x})$, $W(e)$ is independently and uniformly distributed in $[a_e - \delta_e, a_e + \delta_e]$, then $E[W(\mathrm{supp}(\boldsymbol{x}))] \leq B$ holds.*

Thus, we obtain a condition derived as a relaxation of the chance constraint.

Lemma 3. [21] *The function* $f : (k+1)^V \mapsto \mathbb{R}_+$ *is* k*-submodular. For any* $\boldsymbol{x}, \boldsymbol{y} \in (k+1)^V$ *with* $\boldsymbol{x} \preceq \boldsymbol{y}$*, we have*

$$f(\boldsymbol{y}) - f(\boldsymbol{x}) \leq \sum_{e \in \operatorname{supp}(\boldsymbol{y}) \setminus \operatorname{supp}(\boldsymbol{x})} \Delta_{e,\boldsymbol{y}(e)} f(\boldsymbol{x}).$$

3 Uniform IID Weights

In this section, we consider the case where all element weights are independently and identically distributed following a uniform distribution over $[a-\delta, a+\delta]$, where $0 \leq \delta \leq a$. We now give a lower bound on $|\operatorname{supp}(\boldsymbol{x})|$ using the surrogate function based on the Chernoff bound. Since all items have the same expected weight a, we have $E[W(\operatorname{supp}(\boldsymbol{x}))] = a|\operatorname{supp}(\boldsymbol{x})|$. Using Corollary 1, the chance constraint is met if $B - a|\operatorname{supp}(\boldsymbol{x})| \geq \sqrt{3\delta|\operatorname{supp}(\boldsymbol{x})|\ln(1/\alpha)}$. So any $\boldsymbol{x}$ with $|\operatorname{supp}(\boldsymbol{x})| + \frac{\sqrt{3\delta|\operatorname{supp}(\boldsymbol{x})|\ln(1/\alpha)}}{a} \leq \lfloor \frac{B}{a} \rfloor$ is feasible when using the surrogate function. Let $p^* = \max\left\{|\operatorname{supp}(\boldsymbol{x})| : \frac{\sqrt{3\delta|\operatorname{supp}(\boldsymbol{x})|\ln(1/\alpha)}}{a} + |\operatorname{supp}(\boldsymbol{x})| \leq \lfloor \frac{B}{a} \rfloor\right\}$.

So we express the original problem with the chance constraint as Problem 1: maximize $f(\boldsymbol{x})$ subject to $\Pr[W(\operatorname{supp}(\boldsymbol{x})) > B] \leq \alpha$, let Problem 2 be maximizing $f(\boldsymbol{x})$ subject to $|\operatorname{supp}(\boldsymbol{x})| \leq p^*$, Problem 3 is defined as maximizing $f(\boldsymbol{x})$ subject to $|\operatorname{supp}(\boldsymbol{x})| \leq \lfloor B/a \rfloor$. In all three problems, $\boldsymbol{x}$ satisfies $\boldsymbol{x} \in (k+1)^V$, and their relationships are as follows.

Our goal is to design an algorithm whose value of the output solution is as close as possible to the optimal value of Problem 1. Since the k-set that satisfies the constraint of Problem 2 is feasible, we will design an algorithm for Problem 2. By Lemma 2, we know that if a k-set is a feasible solution to Problem 1, then it must be a feasible solution to Problem 3. The approximation ratio refers to the ratio between the function value of a feasible solution to Problem 1 and its optimal value, which is greater than or equal to the ratio between the function value of the solution output by the algorithm designed for Problem 2 and the optimal value of Problem 3. Consequently, we can obtain the approximation ratio of the proposed algorithm.

However, before proposing the decreasing-threshold procedure (given as Algorithm 3) for the problem we studied, we will first propose Algorithm 1 to estimate the range of the optimal value $f(\boldsymbol{o})$ of Problem 3, so as to design the threshold and reduce the query complexity of Algorithm 3. Algorithm 1 initializes the candidate solution $\boldsymbol{s}$ as the zero vector. For each incoming element e, it identifies the optimal position i_e among the k possible sets that maximizes the marginal gain. If the condition $\Delta_{e,i_e} f(\boldsymbol{s}) \geq \frac{f(\boldsymbol{s})}{\lfloor B/a \rfloor}$ is satisfied, the tuple (e, i_e) is added to $\boldsymbol{s}$. The above procedure constitutes the complete Algorithm 1.

Let $\boldsymbol{o}$ be the optimal solution to Problem 3, $\boldsymbol{s}$ the solution returned by Algorithm 1 containing t tuples, $\boldsymbol{s}^j$ the first j tuples of $\boldsymbol{s}$, and $\boldsymbol{o}^j = (\boldsymbol{o} \sqcup \boldsymbol{s}^j) \sqcup \boldsymbol{s}^j$.

Lemma 4. *For all* $0 \leq j \leq t$*, it holds that* $f(\boldsymbol{o}) - f(\boldsymbol{o}^j) \leq f(\boldsymbol{s}^j)$.

Next, we utilize the results on the function value relationships of the constructed intermediate variables to establish the relationship between the function value of the solution returned by Algorithm 1 and that of the optimal solution.

Lemma 5. *From Algorithm* 1, *we can get* $f(\boldsymbol{s}) > f(\boldsymbol{o})/3$.

We know that the solution of Problem 3 should not contain more than $\lfloor B/a \rfloor$ number of tuples, so we choose the last $\lfloor B/a \rfloor$ tuples added to $\boldsymbol{s}$ as its solution and analyze its function value with respect to the function value of $\boldsymbol{s}$.

Lemma 6. *Let* $\boldsymbol{s}_{\lfloor \frac{B}{a} \rfloor}$ *denote the k-set consisting of the last* $\lfloor \frac{B}{a} \rfloor$ *tuples added to* $\boldsymbol{s}$. *Then we have* $f\left(\boldsymbol{s}_{\lfloor \frac{B}{a} \rfloor}\right) \geq f(\boldsymbol{s})/2$.

Proposition 1. *Algorithm* 1 *takes* $O(nk)$ *queries. And if* $\Gamma = \frac{f(\boldsymbol{s})}{2}$, *then* $\Gamma \leq f(\boldsymbol{o}) < 6\Gamma$.

Now we have obtained the estimated range of the optimal value $f(\boldsymbol{o})$, and next we propose a decreasing-threshold procedure algorithm. It employs a decreasing sequence of thresholds based on the estimated range of $f(\boldsymbol{o})$. For each threshold, it examines all item-type tuples such that the item has not yet been selected in any type. For each such tuple, it identifies the position $i_e \in [k]$ that yields the maximum marginal gain. If the maximum marginal gain is greater than or equal to the current threshold and the total number of selected items is less than p^*, the corresponding tuple is added to the solution.

Let $\boldsymbol{o}$ denote an optimal solution of Problem 3 with maximal value, and let $\boldsymbol{s}$ be the solution returned by Algorithm 3. Suppose the algorithm selects l elements in total (with $l \leq p^*$) through iterative point selection. We construct a sequence $\boldsymbol{s}^0, \boldsymbol{s}^1, \ldots, \boldsymbol{s}^l$ as follows: Initialize $\boldsymbol{s}^0 = \mathbf{0}$, and let (e_j, i_j) denote the jth selected element and its assigned position. Then, for each $j \in [l]$, define $\boldsymbol{s}^j = \boldsymbol{s}^{j-1} \sqcup (e_j, i_j)$. Clearly, the final solution satisfies $\boldsymbol{s} = \boldsymbol{s}^l$. Correspondingly, we iteratively define a sequence $\boldsymbol{o}^0 = \boldsymbol{o}, \boldsymbol{o}^1, \ldots, \boldsymbol{o}^l$ based on the updates to $\boldsymbol{s}^j$. The process of constructing $\boldsymbol{o}^j$ from $\boldsymbol{o}^{j-1}$ is as follows: select an element in $\operatorname{supp}\left(\boldsymbol{o}^{j-1}\right)$ and delete it, then add e_j obtained by Algorithm 3 to the i_jth set of $\boldsymbol{o}^{j-1}$. Let $S^j = \operatorname{supp}(\mathbf{o}^{j-1}) \setminus \operatorname{supp}(\boldsymbol{s}^{j-1})$, and let the deleted element be o_j. Then if $e_j \in S^j$, we set $o_j = e_j$, and set o_j to be an arbitrary element in S^j otherwise.

Claim. $f\left(\boldsymbol{o}^{j-1}\right) - f\left(\boldsymbol{o}^j\right) < \frac{1}{1-\epsilon}\left(f\left(\boldsymbol{s}^j\right) - f\left(\boldsymbol{s}^{j-1}\right)\right), \forall j \in [p^*]$.

Based on the above Claim, we have the relationship between $f\left(\boldsymbol{s}^{p^*}\right)$ and $f\left(\boldsymbol{o}\right)$. Further, by the definition of p^*, we will get the approximation ratio.

Theorem 2. *For any* $e \in V$, $W(e)$ *is independently and uniformly sampled from* $[a-\delta, a+\delta](0 \leq \delta \leq a)$, *and for* $\epsilon \in (0, 1/2)$, $B = \omega(1)$, *Algorithm* 2 *outputs a* $(1/2 - \epsilon)$*-approximate solution using at most* $O(\frac{nk}{\epsilon}\log\frac{1}{\epsilon})$ *function evaluations.*

Algorithm 2. Decreasing-Threshold Procedure under the Chance Constraint

Require: $V, f, k, a > 0, B > 0, p^* > 0, \epsilon > 0$.
Ensure: A solution $\boldsymbol{s}$.

$\boldsymbol{s} \leftarrow \mathbf{0}, \boldsymbol{s}_{\text{Alg1}} \leftarrow$ Result of Algorithm 1, $\Gamma \leftarrow \frac{f(\boldsymbol{s}_{\text{Alg1}})}{2}$
$w \leftarrow \frac{6\Gamma}{\epsilon(\lfloor B/a \rfloor)}$
while $w > \frac{\epsilon\Gamma}{2(\lfloor B/a \rfloor)}$ **do**
  for $e \in \{e \in E | \boldsymbol{s}(e) = 0\}$ **do**
    $i_e \leftarrow \arg\max_{i \in [k]} \Delta_{e,i} f(\boldsymbol{s})$
    if $\Delta_{e,i_e} f(\boldsymbol{s}) \geq w$ and $|\text{supp}(\boldsymbol{s})| < p^*$ **then**
      $\boldsymbol{s}(e) \leftarrow i_e$
    end if
  end for
  $w = w(1 - \epsilon)$
end while
return $\boldsymbol{s}$

4 Uniform Weights with the Same Dispersion

We now consider items with different random weights $W(e) \in [a_e - \delta, a_e + \delta]$, assuming that they are chosen independently and uniformly at random and that they share a common dispersion δ. Let $a_{\min} = \min_{e \in V} a_e \geq 1$. Moreover, it is assumed that $B = \omega(1)$ increases as the size of the input increases.

The initial issue is shown as Problem 1 below, that is, the goal is to maximize $f(\boldsymbol{x})$ subject to $\Pr[W(\text{supp}(\boldsymbol{x})) > B] \leq \alpha$. We know that any solution $\boldsymbol{x}$ with $E[W(\text{supp}(\boldsymbol{x}))] \leq B - \sqrt{3\delta |\text{supp}(\boldsymbol{x})| \ln(1/\alpha)}$ is feasible when using the surrogate function, since $E[W(\text{supp}(\boldsymbol{x}))] \leq B$ and $a_{\min} \geq 1$, so $|\text{supp}(\boldsymbol{x})| \leq B$. That is, if $E[W(\text{supp}(\boldsymbol{x}))] \leq B - \sqrt{3\delta B \ln(1/\alpha)}$, then $E[W(\text{supp}(\boldsymbol{x}))] \leq B - \sqrt{3\delta |\text{supp}(\boldsymbol{x})| \ln(1/\alpha)}$. So Problem 2 is defined as maximizing $f(\boldsymbol{x})$ subject to $E[W(\text{supp}(\boldsymbol{x}))] \leq B - \sqrt{3\delta B \ln(1/\alpha)}$. And we define Problem 3 as maximizing $f(\boldsymbol{x})$ subject to $E[W(\text{supp}(\boldsymbol{x}))] \leq B$. In all three problems, $\boldsymbol{x}$ satisfies $\boldsymbol{x} \in (k+1)^V$, and their relationships are as follows.

Similar to the previous case, this section will design algorithms for Problem 2, and the optimal value of Problem 3 may be larger than the optimal value of Problem 1. The approximation ratio refers to the ratio of a feasible solution's function value in Problem 1 to its optimum, which is no less than the ratio of the function value of the solution output by the algorithms for Problem 2 to the optimal value of Problem 3. Thus, we establish the approximation guarantee of the proposed algorithms. Next, we will propose Algorithm 3 and then design thresholds based on its results to propose Algorithm 4.

Algorithm 3 partitions the ground set V into two subsets: elements with expected weights not exceeding $(B - 2\sqrt{3\delta B \ln(1/\alpha)})/2$ form the first subset, denoted as V_1, and the remaining elements constitute the second, denoted as V_2. Specifically, Algorithm 3 initializes a candidate solution $\boldsymbol{s} \leftarrow \mathbf{0}$ and a tuple $(e_m, i_m) \leftarrow (\emptyset, 1)$, which keeps track of the best single element assignment in terms of $f((e, i))$. For each incoming element e, the algorithm identifies the position $i_e \in [k]$ that maximizes $f((e,i))$, and the tuple (e_m, i_m) is updated if $f((e,i_e)) > f((e_m,i_m))$. If $E[W(e)] \leq (B - 2\sqrt{3\delta B \ln(1/\alpha)})/2$, it determines the position $i'_e \in [k]$ that maximizes the marginal gain $\Delta_{e,i'_e} f(\boldsymbol{s})$. If the gain satisfies $\Delta_{e,i'_e} f(\boldsymbol{s}) \geq \frac{E[W(e)]f(\boldsymbol{s})}{B}$, the pair (e, i'_e) is added to $\boldsymbol{s}$. After processing all elements, Algorithm 3 selects a refined solution $\boldsymbol{s}'$ as a subset consisting of the last u tuples added to $\boldsymbol{s}$ (where $|\operatorname{supp}(\boldsymbol{s})| = t$), such that their total expected weight is closest to $B - \sqrt{3\delta B \ln(1/\alpha)}$. Finally, the algorithm returns the better of (e_m, i_m) and $\boldsymbol{s}'$ as the final solution $\boldsymbol{s}_{\text{final}}$. The above procedure constitutes the complete Algorithm 3.

The proof idea for the approximation ratio of Algorithm 3 is essentially the same as that for Algorithm 1, with the additional need to analyze the relationship between the optimal value of Problem 3 restricted to V_2 and the function value of Algorithm 3's output solution. Here, we use $\boldsymbol{s}^j$ to denote the first j tuples of $\boldsymbol{s}$.

Lemma 7. *Let $\boldsymbol{o}_1$ be an optimal solution to Problem 3 restricted to V_1, then $f(\boldsymbol{s}^t) > \frac{1}{3} f(\boldsymbol{o}_1)$ holds.*

Next, we will show the relationship between the function value of the last u tuples added into $\boldsymbol{s}$, denoted as $\boldsymbol{s}_u$, and the function value of $\boldsymbol{s}$. Let $v = t - u$, then $\boldsymbol{s}^t = \boldsymbol{s}^v \sqcup \boldsymbol{s}_u$.

Lemma 8. $f(\boldsymbol{s}_u) > f(\boldsymbol{s}^t)/3$.

If there are three elements in the solution of Problem 3 on V_2, then the sum of their expected weights exceeds $\frac{3}{2}B - 3\sqrt{3\delta B \ln(1/\alpha)} > B$ for $B = \omega(1)$. Therefore, there can be at most two elements in the solution on V_2. Let $\boldsymbol{o}_2$ denote an optimal solution of Problem 3 over V_2. It then follows that $f(\boldsymbol{o}_2) \leq 2f((e_m, i_m)) \leq 2f(\boldsymbol{s}_{\text{final}})$. Combining the above argument, we obtain the following theorem.

Theorem 3. *Algorithm 3 takes $O(nk)$ queries and achieves a 1/11-approximation.*

We next propose Algorithm 4, where we use the notation $\boldsymbol{s}^j$ to denote the k-set $\boldsymbol{s}$ after adding j tuples(with $\boldsymbol{s}^0 = \mathbf{0}$), $\boldsymbol{o}$ to denote the optimal solution of Problem 3 on V, $T = \max\left\{j \in \mathbb{N} : \boldsymbol{s}^j + E[W(o_1)] \leq B - \sqrt{3\delta B \ln(1/\alpha)}\right\}$, and $o_1 = \arg\max_{o \in \operatorname{supp}(\boldsymbol{o})} E[W(o)]$. Algorithm 4 consists of two main phases. The first phase aims to generate a candidate solution $\boldsymbol{s}$, while the second phase post-processes the candidate solution to improve the quality of the solution. In the first stage, the threshold is determined according to the range of the optimal value obtained by Algorithm 3. Then, for each threshold, we traverse all elements that have not yet been included in the candidate solution. For each element, a position that maximizes the marginal gain is identified. If its inclusion does not exceed the preset weight limit and its density gain meets the corresponding threshold condition, the element will be added to the candidate solution. The second part improves the quality of the solution by improving the quality of

$\boldsymbol{s}^T$. Since $\boldsymbol{s}^T$ is unknown, $\boldsymbol{s}^T$ is approximated by multiple guesses of the first q elements of $\boldsymbol{s}$, where the expected weight of $\boldsymbol{s}^q$ is the maximum value that does not exceed $\varepsilon(1+\varepsilon)^l(B-\sqrt{3\delta B\ln(1/\alpha)})$, l ranges from 0 to $\lfloor \log_{1+\varepsilon}\varepsilon^{-1}\rfloor$. Then traverse the elements that are not added to $\boldsymbol{s}^q$, find an element that meets the weight condition and produces the maximum marginal gain, and add it to $\boldsymbol{s}^q$.

Algorithm 4. The Threshold Greedy Algorithm under the Chance Constraint

Input: V, f, k, δ, α, $B>0$, $\epsilon>0$.
Output: A solution $\boldsymbol{s}$.
`// Phase 1: Finding candidate solutions`
$\boldsymbol{s}_{\text{Alg3}} \leftarrow$ Result of Algorithm 3, $\boldsymbol{s}\leftarrow \mathbf{0}$
$\Gamma \leftarrow f(\boldsymbol{s}_{\text{Alg3}})$, $\theta \leftarrow 11\Gamma/(3\epsilon(1-\epsilon)(B+\sqrt{3\delta B\ln(1/\alpha)}))$
while $\theta \geq \epsilon(1-\epsilon)\Gamma/(3(B+\sqrt{3\delta B\ln(1/\alpha)}))$ **do**
  for $e\in V\setminus \text{supp}(\boldsymbol{s})$ **do**
    $i_e \leftarrow \arg\max_{i\in[k]} \Delta_{e,i}f(\boldsymbol{s})$
    if $E[W(\text{supp}(\boldsymbol{s}))]+E[W(e)] \leq B-\sqrt{3\delta B\ln(1/\alpha)}$ and $\Delta_{e,i_e}f(\boldsymbol{s})/E[W(e)] \geq \theta$ **then**
      $\boldsymbol{s}\leftarrow \boldsymbol{s}\sqcup(e,i_e)$
    end if
  end for
  $\theta \leftarrow (1-\epsilon)\theta$
end while`// Phase 2: Improving quality of solutions`
for $l=0$ to $\lfloor \log_{1+\varepsilon}\varepsilon^{-1}\rfloor$ **do**
  Let $\boldsymbol{s}_l$ be the k-set $\boldsymbol{s}^q$ for the maximal q value for which $E[W(\text{supp}(\boldsymbol{s}^q))] \leq \varepsilon(1+\varepsilon)^l(B-\sqrt{3\delta B\ln(1/\alpha)})$.
  if there exists an element e_l such that $E[W(e_l)] \leq B-\sqrt{3\delta B\ln(1/\alpha)} - E[W(\text{supp}(\boldsymbol{s}_l))]$ **then**
    $i_l \leftarrow \arg\max_{i\in[k]} \Delta_{e_l,i}f(\boldsymbol{s})$
    Let $(\tilde{e}_l,\tilde{i}_l)$ be the tuple maximizing $f((\tilde{e}_l,\tilde{i}_l)\mid \boldsymbol{s}_l)$ among all the tuples (e_l,i_l) obeying the conditions on the previous two lines.
    Let $\boldsymbol{s}_{(l)}\leftarrow \boldsymbol{s}_l\sqcup(\tilde{e}_l,\tilde{i}_l)$.
  else
    Let $\boldsymbol{s}_{(l)}\leftarrow \boldsymbol{s}_l$.
  end if
end for
$\boldsymbol{s}\leftarrow \arg\max_{\boldsymbol{s}''\in\{\boldsymbol{s}_{\text{Alg3}},\boldsymbol{s},\boldsymbol{s}_{(0)},\ldots,\boldsymbol{s}_{(\lfloor \log_{1+\varepsilon}\varepsilon^{-1}\rfloor)}\}} f(\boldsymbol{s}'')$
return $\boldsymbol{s}$

Lemma 9. *When* $E[W(o_1)] > (1-\epsilon)(B-\sqrt{3\delta B\ln(1/\alpha)})$, *for* $B=\omega(1)$, *Algorithm 4 achieves a* $(1/3-\epsilon)$*-approximation.*

Lemma 10. *When* $E[W(o_1)] \leq (1-\epsilon)(B-\sqrt{3\delta B\ln(1/\alpha)})$, *for* $B=\omega(1)$, *Algorithm 4 achieves a* $(1/3-\epsilon)$*-approximation.*

Combining the last two lemmas, we can obtain an approximation ratio for Algorithm 4, which further yields the following theorem.

Theorem 4. *For any $e \in V$, let $W(e) \in [a_e - \delta, a_e + \delta]$ with $0 \leq \delta \leq \min_{e \in V} a_e$ and $\min_{e \in V} a_e \geq 1$. For $\epsilon \in (0, 1/3)$, $B = \omega(1)$, Algorithm 4 has $O(\frac{kn}{\epsilon} \log \frac{1}{\epsilon})$ query complexity and achieves a $(1/3 - \epsilon)$-approximation.*

5 Conclusion

In this paper, we studied the maximization of monotone k-submodular functions under chance constraints, where random element weights are independently sampled from two types of uniform distributions. We proposed algorithms for both settings that achieve approximation guarantees of $(1/2 - \epsilon)$ and $(1/3 - \epsilon)$, respectively, with provable bounds on query complexity. These results provide theoretical guarantees for chance-constrained k-submodular maximization and may be extended to handle more general weight distributions.

Acknowledgments. This work was supported in part by the National Natural Science Foundation of China (No. 12471306), the Natural Science Foundation of Shandong Province of China (No. ZR2024MA031), and the Fundamental Research Funds for the Central Universities.

Disclosure of Interests. The authors have no competing interests.

References

1. Badanidiyuru, A., Vondrák, J.: Fast algorithms for maximizing submodular functions. In: Proceedings of the Twenty-Fifth Annual ACM-SIAM Symposium on Discrete Algorithms, pp. 1497–1514. SIAM, Portland, Oregon, USA (2014)
2. Calinescu, G., Chekuri, C., Pal, M., Vondrák, J.: Maximizing a monotone submodular function subject to a matroid constraint. SIAM J. Comput. **40**(6), 1740–1766 (2011)
3. Charikar, M., Khuller, S., Mount, D.M., Narasimhan, G.: Algorithms for facility location problems with outliers. In: SODA, pp. 642–651. Society for Industrial and Applied Mathematics, Washington, D.C., USA (2001)
4. Chen, J., Maehara, T.: Chance-constrained submodular knapsack problem. In: Du, D., Duan, Z., Tian, C. (eds.) Computing and Combinatorics. LNCS, vol. 11653, pp. 103–114. Springer, Heidelberg (2019). https://doi.org/10.1007/978-3-030-26176-4_9
5. Doerr, B., Doerr, C., Neumann, A., Neumann, F., Sutton, A.: Optimization of chance-constrained submodular functions. In: Proceedings of the AAAI Conference on Artificial Intelligence, pp. 1460–1467. AAAI Press, New York, New York, USA (2020)
6. Feige, U.: A threshold of ln n for approximating set cover. J. ACM (JACM) **45**(4), 634–652 (1998)
7. Gong, S., Liu, B., Fang, Q.: Streaming submodular maximization with the chance constraint. In: Li, M., Sun, X. (eds.) Frontiers of Algorithmic Wisdom. LNCS, vol. 13461, pp. 129–140. Springer, Heidelberg (2022). https://doi.org/10.1007/978-3-031-20796-9_10

8. Ha, D.T., Pham, C.V., Tran, T.D.: Improved approximation algorithms for k-submodular maximization under a knapsack constraint. Comput. Oper. Res. **161**, 106452 (2024)
9. Kulik, A., Shachnai, H., Tamir, T.: Approximations for monotone and nonmonotone submodular maximization with knapsack constraints. Math. Oper. Res. **38**(4), 729–739 (2013)
10. Lee, J., Sviridenko, M., Vondrák, J.: Submodular maximization over multiple matroids via generalized exchange properties. Math. Oper. Res. **35**(4), 795–806 (2010)
11. Lin, H., Bilmes, J.: Multi-document summarization via budgeted maximization of submodular functions. In: Human Language Technologies: The 2010 Annual Conference of the North American Chapter of the Association for Computational Linguistics, pp. 912–920. Association for Computational Linguistics, Los Angeles, California (2010)
12. Nemhauser, G.L., Wolsey, L.A., Fisher, M.L.: An analysis of approximations for maximizing submodular set functions–I. Math. Program. **14**(1), 265–294 (1978)
13. Neumann, A., Neumann, F.: Optimizing monotone chance-constrained submodular functions using evolutionary multiobjective algorithms. In: Evolutionary Computation, pp. 1–31 (2024)
14. Nie, G., Zhu, Y., Nadew, Y.Y., Basu, S., Pavan, A., Quinn, C.J.: Size-constrained k-submodular maximization in near-linear time. In: Uncertainty in Artificial Intelligence, pp. 1545–1554. PMLR, Pittsburgh, PA, USA (2023)
15. Ohsaka, N., Yoshida, Y.: Monotone k-submodular function maximization with size constraints. Adv. Neural. Inf. Process. Syst. **28**, 694–702 (2015)
16. Qian, C., Shi, J.C., Tang, K., Zhou, Z.H.: Constrained monotone k-submodular function maximization using multiobjective evolutionary algorithms with theoretical guarantee. IEEE Trans. Evol. Comput. **22**(4), 595–608 (2017)
17. Queyranne, M.: Minimizing symmetric submodular functions. Math. Programm. **82**(1), 3–12 (1998)
18. Sviridenko, M.: A note on maximizing a submodular set function subject to a knapsack constraint. Oper. Res. Lett. **32**(1), 41–43 (2004)
19. Tang, Z., Chen, J., Wang, C.: Greedy+ singleton: an efficient approximation algorithm for k-submodular knapsack maximization. Theoret. Comput. Sci. **984**, 114320 (2024)
20. Tang, Z., Wang, C., Chan, H.: On maximizing a monotone k-submodular function under a knapsack constraint. Oper. Res. Lett. **50**(1), 28–31 (2022)
21. Xiao, H., Liu, Q., Zhou, Y., Li, M.: Approximation algorithms for k-submodular maximization subject to a knapsack constraint. J. Oper. Res. Soc. China, 1–16 (2024). https://doi.org/10.1007/s40305-024-00539-y
22. Yan, X., Do, A.V., Shi, F., Qin, X., Neumann, F.: Optimizing chance-constrained submodular problems with variable uncertainties. In: ECAI 2023, pp. 2826–2833. IOS Press, Kraków, Poland (2023)
23. Yan, X., Neumann, A., Neumann, F.: Sliding window bi-objective evolutionary algorithms for optimizing chance-constrained monotone submodular functions. In: Affenzeller, M., et al. (eds.) International Conference on Parallel Problem Solving from Nature, pp. 20–35. Springer, Hagenberg, Austria (2024). https://doi.org/10.1007/978-3-031-70055-2_2

A Modified Parallel Tangent Global Search Algorithm for Finding Anti-Berge Equilibrium in Bimatrix Games

Alexander Gornov[1], Enkhbat Rentsen[2], Pavel Sorokovikov[1(✉)], and Tatiana Zarodnyuk[1]

[1] Matrosov Institute for System Dynamics and Control Theory of Siberian Branch of Russian Academy of Sciences, Lermontov Street, 134, 664033 Irkutsk, Russia
{pavel,tz}@icc.ru,enkhbatm@num.edu.mn

[2] The Institute of Mathematics and Digital Technology, Mongolian Academy of Sciences, Ulaanbaatar, Mongolia
enkhbatm@num.edu.mn
http://idstu.irk.ru/en

Abstract. This study addresses the problem of finding an anti-Berge equilibrium in a bimatrix game based on a global search algorithm. Finding the anti-Berge equilibrium reduces equivalently to a quadratic programming with an indefinite matrix and linear constraints, which belongs to a class of global optimization. To solve the problem numerically, we develop a modified parallel tangent algorithm. The proposed algorithm uses the one-dimensional nonlocal search procedure based on the Strongin and parabolas methods. The stop criterion of the algorithm is the sufficient condition for anti-Berge equilibrium. The proposed algorithm is implemented and numerically tested on a range of bimatrix games.

Keywords: Game theory · Anti-Berge equilibrium · Global optimization · Parallel tangent algorithm

1 Introduction

Game theory is widely recognized for its crucial contribution to optimization and operations research, mathematical modeling, applied mathematics, decision theory, and economics. The field has been fundamentally shaped by the study of solution concepts such as the Nash equilibrium [1–5], which describes a state where no player can benefit by unilaterally changing their strategy, and the Berge equilibrium [6–12], which embodies a cooperative principle where players maximize each other's payoffs. While these concepts have been extensively explored in the context of both zero-sum and nonzero-sum games (see, for example, [13–35]), the continuous development of game theory necessitates the introduction of new equilibria types to model more complex strategic interactions, including scenarios of antagonism or sabotage.

A. Chinchuluun et al. (Eds.): AAIM 2025, LNCS 16379, pp. 59–72, 2026.
https://doi.org/10.1007/978-981-95-5657-1_6

A concept of a new equilibrium called anti-Berge has recently appeared [36]. In contrast to the Berge equilibrium, where players seek to maximize their opponents' payoffs, an anti-Berge equilibrium describes a state in which each player aims to minimize the payoff of their opponents. This concept is particularly relevant for modeling conflictual relationships. However, to the best of our knowledge, no general methods or algorithms for computing this equilibrium have been proposed yet.

This study aims to fill this gap by developing a global optimization algorithm for finding anti-Berge equilibria. The paper is organized as follows. Section 2 formulates the problem of finding an anti-Berge equilibrium in a bimatrix game as a global optimization problem with linear constraints. Section 3 describes the proposed global search algorithm. Section 4 presents numerical experiments applied to several bimatrix games, demonstrating the effectiveness of our approach.

2 Bimatrix Game

Consider the bimatrix game in mixed strategies with matrices (A, B) for players 1 and 2.

$$A = (a_{ij}),\ i = 1, \dots, m,$$
$$B = (b_{ij}),\ j = 1, \dots, n.$$

The payoff entries a_{ij} and b_{ij} are real numbers. Denote by X and Y the sets

$$X = \{x \in R^m \mid \sum_{i=1}^{m} x_i = 1,\ x_i \geq 0,\ i = 1, \dots, m\},$$

$$Y = \{y \in R^n \mid \sum_{j=1}^{n} y_j = 1,\ y_j \geq 0,\ j = 1, \dots, n\}.$$

The mixed strategy for player 1 is vector $x = (x_1, x_2, \dots, x_m)^T \in X$ representing the probability that player 1 uses strategy x^i. Similarly, the mixed strategy for player 2 is $y = (y_1, y_2, \dots, y_n)^T \in Y$. Their expected payoffs are determined as follows:

$$f_1(x, y) = x^T Ay, \qquad f_2(x, y) = x^T By.$$

Let us recall the following definitions and theorems on anti-Berge equilibrium.

Definition 1. *[36] A pair strategy $(x^1, y^1) \in X \times Y$ is an anti-Berge equilibrium (with respect to player 2) if*

$$\begin{cases} f_1(x^1, y^1) \geq f_1(x^1, y), & \forall y \in Y, \\ f_2(x^1, y^1) \leq f_2(x, y^1), & \forall x \in X. \end{cases}$$

It is clear that

$$f_1(x^1, y^1) = \max_{y \in Y} f_1(x^1, y),$$

$$f_2(x^1, y^1) = \min_{x \in X} f_2(x, y^1).$$

Definition 2. *[36] A pair strategy* $(x^2, y^2) \in X \times Y$ *is an anti-Berge equilibrium (with respect to player 1) if*

$$\begin{cases} f_1(x^2, y^2) \leq f_1(x^2, y), & \forall y \in Y, \\ f_2(x^2, y^2) \geq f_2(x, y^2), & \forall x \in X. \end{cases}$$

Theorem 1. *[36] A pair strategy* (x^*, y^*) *is an anti-Berge equilibrium (with respect to player 2) for the bimatrix game if and only if there exist scalars* (p^*, q^*) *such that* (x^*, y^*, p^*, q^*) *is a solution to the following quadratic programming problem:*

$$\max_{(x,y,p,q)} F(x, y, p, q) = x^T (A - B) y - p + q \tag{1}$$

subject to:

$$\left[x^T A\right]_j \leq p, \ j = 1, \ldots, n, \tag{2}$$

$$[By]_i \geq q, \ i = 1, \ldots, m, \tag{3}$$

$$\sum_{i=1}^{m} x_i = 1, \ x_i \geq 0, \ i = 1, \ldots, m, \tag{4}$$

$$\sum_{j=1}^{n} y_j = 1, \ y_j \geq 0, \ j = 1, \ldots, n. \tag{5}$$

We can also formulate the following assertion for anti-Berge equilibrium (with respect to player 1).

Theorem 2. *[36] A pair strategy* $(\hat{x}^*, \hat{y}^*)$ *is an anti-Berge equilibrium (with respect to player 1) for the bimatrix game if and only if there exist scalars* $(\hat{p}^*, \hat{q}^*)$ *such that* $(\hat{x}^*, \hat{y}^*, \hat{p}^*, \hat{q}^*)$ *is a solution to the following quadratic programming problem:*

$$\max_{(x,y,p,q)} F(x, y, p, q) = x^T (B - A) y + p - q$$

subject to:

$$\left[x^T A\right]_j \geq p, \ j = 1, \ldots, n,$$

$$[By]_i \leq q, \ i = 1, \ldots, m,$$

$$\sum_{i=1}^{m} x_i = 1, \ x_i \geq 0, \ i = 1, \ldots, m,$$

$$\sum_{j=1}^{n} y_j = 1, \ y_j \geq 0, \ j = 1, \ldots, n.$$

Note that the condition

$$F(x^*, y^*, p^*, q^*) = 0 \tag{6}$$

is necessary and sufficient for (x^*, y^*) to be an anti-Berge equilibrium, which will be used in the proposed global search algorithm.

3 The Computational Technique for Finding Anti-Berge Equilibrium

To solve the above problem, we use the modification of the Parallel Tangent Algorithm (Partan). The Parallel Tangent Algorithm [37] is an optimization procedure in which the conjugated trajectories are constructed by solving one-dimensional problems. We propose a globalized version of the parallel tangent method (Subsect. 3.1), which incorporates the univariate global search procedure (Subsect. 3.2) as an auxiliary one (steps 2, 4, 6 of the Partan algorithm). The one-dimensional nonlocal search scheme developed is a hybrid approach based on the Strongin method [38] and parabolic technique.

The computational technique devised relies on deterministic algorithms whose convergence is strictly proven [39,40]. On their basis, by adding globalizing procedures and building new combinations of algorithms, an approach was developed to find the anti-Berge equilibrium.

The parallel tangent programming algorithm was initially designed to solve box-constrained optimization problems. Therefore, we use penalty function techniques to transform our problem from a constrained to an unconstrained one. We switch from four variables x, y, p, q to one aggregate vector $w = (x, y, p, q)$, in which vector variables x, y and scalar variables p, q are combined. A single penalty function $\hat{g}(w) = \tilde{g}^2(w)$ is created for each equality constraint $\tilde{g}(w) = 0$. The relevant penalty function is specified for each inequality constraint $\tilde{q}(w) \leq 0$:

$$\hat{q}(w) = \begin{cases} 0, & \text{if } \tilde{q}(w) \leq 0, \\ \tilde{q}^2(w), & \text{if } \tilde{q}(w) > 0. \end{cases} \tag{7}$$

Let us consider a box-bounded minimization:

$$\hat{f}(w) = F(w) + \frac{\gamma}{2}\sum_i \hat{g}_i(w) + \frac{\gamma}{2}\sum_j \hat{q}_j(w) \to \min_W, \tag{8}$$

$$W = \{w \in \mathbb{R}^s : \underline{w}_i \leq w_i \leq \overline{w}_i,\ i = 1, \ldots, n+m+2\}, \tag{9}$$

$$s = n + m + 2, \tag{10}$$

$$\underline{w}_i = 0, i = 1, \ldots, n+m+2, \overline{w}_j = 1, j = 1, \ldots, n+m, \tag{11}$$

$$\overline{w}_{n+m+1} = \overline{p}, \overline{w}_{n+m+2} = \overline{q}, \tag{12}$$

$$\hat{g}_j(w) = \sum_{j=1}^{n+m} w_j - 1, \quad j = 1, \ldots, n+m, \tag{13}$$

$$w_i = x_i, \quad i = 1, \ldots, m,\ w_{m+j} = y_j, \quad j = 1, \ldots, n, \tag{14}$$

$$w_{n+m+1} = p, \quad w_{n+m+2} = q, \tag{15}$$

$$\hat{q}_j(w) = \left[w^T A\right]_j - w_{n+m+1},\ j = 1, \ldots, n, \tag{16}$$

$$\hat{q}_{n+i}(w) = w_{n+m+2} - \left[By\right]_i,\ i = 1, \ldots, m. \tag{17}$$

In this case, the coefficient $\gamma \in \mathbb{R}$ is used to penalize, $\overline{p} \in \mathbb{R}$, $\overline{q} \in \mathbb{R}$ are constants whose values are chosen from certain ranges.

The selection of key parameters in the proposed algorithm is driven by a balance between computational efficiency and global convergence guarantees. The initial coefficient value γ is not very large (e.g. $\gamma = 10^3$) and after some local optima are found, it rises for searching another local extremum. The initial penalty coefficient γ and its update rule are chosen to gradually enforce constraints without causing immediate ill-conditioning of the penalized function. Following tradition, the value of the penalty increases consistently during the optimization process. The accuracy of the solution obtained can be enhanced by its iterative growth. The accuracy parameters ε_f and ε_λ are set based on the desired solution precision and the scales of the payoff matrices in the problem. The safety factor ν_s in the one-dimensional search follows established practices from the Strongin algorithm to ensure reliable estimation of the Lipschitz constant.

3.1 Modification of the Parallel Tangent Algorithm

Below, the authors' modification of the Partan method for finding the bimatrix game equilibrium is presented.

Step 1. Select the starting point $w^0 \in W$, the accuracy of the function value $\varepsilon_f \in (0,\ 1) \subset \mathbb{R}$. The value of ε_f is chosen relative to the expected scale of the payoff function $F(w)$ to ensure meaningful convergence.

Step 2. Perform a one-dimensional minimization of function $\hat{f}$ along the direction $d^s = e^s = (0,\ ...,\ 0,\ 1)^T$. Find the step size λ_0 from the condition

$$\hat{f}\left(w^0 + \lambda_0 d^s\right) = \min_{\lambda} \hat{f}\left(w^0 + \lambda d^s\right). \tag{18}$$

Set $w^1 = w^0 + \lambda_0 d^s$ and specify the number of iteration $k = 1$.

Step 3. Take the directions of the axes system as the starting directions for the optimization, i.e. $d^i = e^i$, $i = \overline{1, s}$, where $e^1 = (1,\ ...,\ 0,\ 0)^T$, ..., $e^s = (0,\ ...,\ 0,\ 1)^T$.

Step 4. Perform s univariate searches from the point w^0 along directions $d^i = e^i$, $i = \overline{1, s}$. Each subsequent search is made from the approximation obtained in the previous step. Obtain the step size λ_i from the condition

$$\hat{f}\left(w^{i-1} + \lambda_i d^i\right) = \min_{\lambda} \hat{f}\left(w^{i-1} + \lambda d^i\right). \tag{19}$$

Set $w^i = w^{i-1} + \lambda_i d^i$.

Step 5. Take a new direction $d = w^s - w^0$. Change directions $d^1, ..., d^s$ by $d^2, ..., d^s, d$, accordingly.

Step 6. Perform a one-dimensional minimization from point w^s along the direction $d = d^s = w^s - w^0$. Obtain the step size λ_{s+1} from the condition

$$\hat{f}\left(w^s + \lambda_{s+1} d\right) = \min_{\lambda} \hat{f}\left(w^s + \lambda d\right). \tag{20}$$

Set $w^{s+1} = w^s + \lambda_{s+1} d$, $k = k + 1$.

Step 7. Check if the condition $k \leq s$ is met. If yes, move to step 8. Otherwise, go to step 9.

Step 8. Check if the termination condition $F(w^s) \leq \varepsilon_f$ (see (2.6)) is reached. If yes, end the search, obtain the final solution $w^* = w^{s+1}$. Otherwise, move to step 9.

Step 9. Assume $w^0 = w^{s+1}$, take this point as the initial approximation w^0 for the next iteration. Double penalty parameter $\gamma = 2 \cdot \gamma$. This multiplicative update strategy provides a balanced increase: it is aggressive enough to eventually enforce constraint satisfaction but gradual enough to avoid premature numerical instability that can occur with very large penalty parameters. Move to step 4.

In order to solve an auxiliary univariate optimization problem, the proposed tangent algorithm implements the one-dimensional global search procedure.

3.2 The One-Dimensional Non-local Optimization Method

We have developed the procedure based on the Strongin global search algorithm and the parabolic method, focusing on the construction of "convex triples". It facilitates finding the conjugate descent directions for the main tangent algorithm. The hybrid one-dimensional procedure alternates between parabolic interpolation (on odd iterations) and the Lipschitz-bound global search (on even iterations). This design leverages the fast local convergence of the parabolic method when a promising region is identified, while the Strongin-style search ensures global explorative capability by systematically investigating intervals with high potential utility.

Step 1. Specify the number of initial trials $\zeta_t \in \mathbb{N}$, the accuracy $\varepsilon_\lambda \in (0,\ 1) \subset \mathbb{R}$, the safety factor $\nu_s \in \mathbb{R}$ $(\nu_s > 1)$. Choose $\lambda_1 = \alpha$, $\lambda_{\zeta_t} = \beta$, $\kappa = \zeta_t$, $N_{it} = 1$. The points λ_j, $2 \leq j \leq \zeta_t - 1$ are randomly distributed at $[\alpha,\ \beta]$. The number of initial trials ζ_t is set to provide a sufficiently diverse initial exploration of the interval. The safety factor ν_s is typically chosen in the range $[1.5, 3.0]$ to provide a robust overestimate of the local Lipschitz constant, preventing the algorithm from overlooking narrow minima.

Step 2. Sort λ_j, $1 \leq j \leq \kappa$:

$$\alpha = \lambda_1 < \lambda_2 < ... < \lambda_\kappa = \beta. \tag{21}$$

Step 3. If the number of iterations N_{it} is an odd integer, then

Step 3.1. Suppose $f_{\max} = -\infty$. Check any triples of the points λ_j, λ_{j+1}, λ_{j+2}, $1 \leq j \leq \zeta_t - 2$. If $\hat{f}(\lambda_j) \geq \hat{f}(\lambda_{j+1}) \leq \hat{f}(\lambda_{j+2})$, and $\hat{f}(\lambda_{j+1}) > f_{\max}$, then $f_{\max} = \hat{f}(\lambda_{j+1})$.

Step 3.2. For each $\tilde{\lambda}_j$, $\tilde{\lambda}_{j+1}$, $\tilde{\lambda}_{j+2}$ compute the assessment $\delta_j = \left| f_{\max} - \hat{f}\left(\tilde{\lambda}_{j+1}\right) \right|$.

Step 3.3. Choose stochastically the number s of the triple from the computed values δ_j, with non-zero probability for each triple.

Step 3.4. Search for a minimum point λ_s^* of the parabola obtained from the selected triple using a local one-dimensional optimization technique. Take $\lambda^{\kappa+1} = \lambda_s^*$, calculate $\hat{f}(\lambda^{\kappa+1})$.

Step 4. If the number of iterations N_{it} is an even integer, then

Step 4.1. Estimate the Lipschitz constant m:

$$M = \max \frac{\left|\hat{f}_j - \hat{f}_{j-1}\right|}{\left|\lambda_j - \lambda_{j-1}\right|}, m = \begin{cases} 1, \ M = 0 \\ M, \ M > 0 \end{cases}, \hat{f}_j = \hat{f}(\lambda_j), 1 \leq j \leq \kappa. \tag{22}$$

Step 4.2. Calculate the characteristic values for each segment $(\lambda_{j-1}, \lambda_j)$, $1 \leq j \leq \kappa$:

$$S(i) = (\lambda_j - \lambda_{j-1}) + \frac{\left(\hat{f}_j - \hat{f}_{j-1}\right)^2}{m^2 {\nu_s}^2 (\lambda_j - \lambda_{j-1})} - 2\frac{\hat{f}_j + \hat{f}_{j-1}}{m\nu_s}. \tag{23}$$

Step 4.3. Select the interval $(\lambda_{\tau-1}, \lambda_\tau)$ in which the maximum of the function

$$S(\tau) = \max\{S(j), \ 1 \leq j \leq \kappa\} \tag{24}$$

is reached.

Step 4.4. If stopping condition

$$\lambda_\tau - \lambda_{\tau-1} \leq \varepsilon_\lambda \tag{25}$$

is satisfied, then end the search.

Otherwise, select the point

$$\lambda^{\kappa+1} = \frac{1}{2}(\lambda_\tau + \lambda_{\tau-1}) - \text{sign}(\hat{f}_\tau - \hat{f}_{\tau-1})\frac{1}{2\nu_s}\left[\frac{\hat{f}_\tau - \hat{f}_{\tau-1}}{m}\right], \tag{26}$$

calculate the value $\hat{f}(\lambda^{\kappa+1})$.

Step 5. Increase $\kappa = \kappa + 1$, $N_{it} = N_{it} + 1$. Go to step 2.

The probabilistic choice of the neighborhoods for the minimum search is one of the traditional approaches in global optimization. After selecting the area that contains the extremum (see paragraph 3.2, step 3 of the algorithm, where the probabilistic selection of neighborhoods occurs), a deterministic search for a solution is used (see paragraph 3.2, step 4). We apply the well-known stopping condition of the Strongin algorithm, the convergence of which is strictly proven [38].

4 Numerical Experiments

The proposed algorithm was implemented in the C language and tested on a range of two-player games using the GCC compiler across the Microsoft Windows, Mac OS X and Linux environments. Several problems were solved using the developed numerical optimization algorithm. The games were constructed with the popular GAMUT package [41]. The computational experiments were carried out on a system featuring an Intel Core i7-1065G7 processor and 16 GB DDR4 RAM operating at 1333 MHz. The proposed algorithm was started from different points uniformly distributed in the search space with set parameter values ($\zeta_t = 4$, $\varepsilon_f = 10^{-7}$, $\varepsilon_\lambda = 10^{-5}$, $\nu_s = 2$). Anti-Berge equilibria were obtained correctly in all cases, each in a few seconds of CPU time. Various equilibrium points in the games were obtained in the course of optimization by successively increasing the values of the penalty parameter γ (see paragraph 3.1).

4.1 Problem 1 ($m = 3, n = 6$)

The first test problem has the smallest dimension among the considered ones (11 variables).

$$A = \begin{pmatrix} 3 & 8 & 14 & 2 & 7 & 19 \\ 5 & 1 & 16 & 4 & 11 & 3 \\ 20 & 6 & 15 & 13 & 10 & 8 \end{pmatrix}, \quad B = \begin{pmatrix} 14 & 1 & 7 & 20 & 12 & 5 \\ 8 & 7 & 2 & 13 & 11 & 9 \\ 3 & 19 & 6 & 10 & 4 & 11 \end{pmatrix}.$$

Each of its four solutions was obtained in less than a second of CPU time. In the process of searching for equilibrium points, the proposed algorithm managed to obtain the exact value of the quality function $F^* = 0.0$ in two out of four cases (see Table 1).

Solutions:

1. $x_1^* = 1.0$, $x_2^* = 0.0$, $x_3^* = 0.0$,
   $y_1^* = 0.0$, $y_2^* = 0.0$, $y_3^* = 0.0$, $y_4^* = 0.0$, $y_5^* = 0.0$, $y_6^* = 1.0$.
2. $x_1^* = 0.0$, $x_2^* = 1.0$, $x_3^* = 0.0$,
   $y_1^* = 0.0$, $y_2^* = 0.0$, $y_3^* = 1.0$, $y_4^* = 0.0$, $y_5^* = 0.0$, $y_6^* = 0.0$.
3. $x_1^* = 0.722223$, $x_2^* = 0.277777$, $x_3^* = 0.0$,
   $y_1^* = 0.0$, $y_2^* = 0.0$, $y_3^* = 0.444441$, $y_4^* = 0.0$, $y_5^* = 0.0$, $y_6^* = 0.555559$.
4. $x_1^* = 0.0$, $x_2^* = 0.312501$, $x_3^* = 0.687499$,
   $y_1^* = 0.444447$, $y_2^* = 0.0$, $y_3^* = 0.555553$, $y_4^* = 0.0$, $y_5^* = 0.0$, $y_6^* = 0.0$.

4.2 Problem 2 ($m = 4, n = 8$)

In this problem, the dimension is larger than in the previous one (14 variables).

$$A = \begin{pmatrix} 5 & 39 & 4 & 7 & 48 & 42 & 16 & 38 \\ 35 & 16 & 8 & 2 & 19 & 22 & 13 & 30 \\ 2 & 41 & 47 & 6 & 1 & 28 & 13 & 21 \\ 8 & 21 & 47 & 1 & 19 & 36 & 15 & 32 \end{pmatrix}, \quad B = \begin{pmatrix} 48 & 13 & 23 & 4 & 4 & 26 & 45 & 47 \\ 27 & 12 & 16 & 7 & 24 & 14 & 18 & 27 \\ 24 & 33 & 3 & 30 & 43 & 26 & 50 & 49 \\ 48 & 25 & 42 & 2 & 41 & 38 & 13 & 28 \end{pmatrix}.$$

Table 1. Problem solutions p^*, q^*, F^*

Problem	Solution	p^*	q^*	F^*
3×6	1	19.0	5.0	0.0
	2	16.0	2.0	0.0
	3	14.555561	5.888883	$2.3 \cdot 10^{-8}$
	4	15.312499	4.666659	$2.8 \cdot 10^{-8}$
4×8	1	48.0	4.0	0.0
	2	34.190452	25.500023	$1.6 \cdot 10^{-8}$
	3	30.285703	14.470610	$2.2 \cdot 10^{-8}$
	4	28.124995	27.000030	$3.1 \cdot 10^{-8}$
	5	47.0	3.0	0.0
5×9	1	44.0	2.0	0.0
	2	38.999972	13.846166	$2.5 \cdot 10^{-8}$
	3	38.999973	8.666690	$3.7 \cdot 10^{-8}$
6×10	1	48.399974	20.448297	0.0
	2	36.764149	12.897908	$1.4 \cdot 10^{-8}$
	3	46.759974	13.571437	0.0
	4	36.991319	15.901180	$1.8 \cdot 10^{-8}$
9×15	1	43.0	4.0	0.0
	2	27.763149	15.692326	$3.6 \cdot 10^{-8}$
	3	47.0	16.0	0.0
	4	49.0	7.0	0.0
	5	40.0	10.600023	0.0

By numerically solving this problem using the developed algorithm, we found five anti-Berge equilibrium points.

Solutions:

1. $x_1^* = 1.0$, $x_2^* = 0.0$, $x_3^* = 0.0$, $x_4^* = 0.0$,
   $y_1^* = 0.0$, $y_2^* = 0.0$, $y_3^* = 0.0$, $y_4^* = 0.0$, $y_5^* = 1.0$, $y_6^* = 0.0$, $y_7^* = 0.0$, $y_8^* = 0.0$.
2. $x_1^* = 0.523810$, $x_2^* = 0.476190$, $x_3^* = 0.0$, $x_4^* = 0.0$,
   $y_1^* = 0.0$, $y_2^* = 0.0$, $y_3^* = 0.0$, $y_4^* = 0.0$, $y_5^* = 0.5$, $y_6^* = 0.0$, $y_7^* = 0.0$, $y_8^* = 0.5$.
3. $x_1^* = 0.0$, $x_2^* = 0.428571$, $x_3^* - 0.571429$, $x_4^* = 0.0$,
   $y_1^* = 0.0$, $y_2^* = 0.382353$, $y_3^* = 0.617647$, $y_4^* = 0.0$,
   $y_5^* = 0.0$, $y_6^* = 0.0$, $y_7^* = 0.0$, $y_8^* = 0.0$.
4. $x_1^* = 0.0$, $x_2^* = 0.791667$, $x_3^* = 0.208333$, $x_4^* = 0.0$,
   $y_1^* = 0.879999$, $y_2^* = 0.0$, $y_3^* = 0.0$, $y_4^* = 0.0$,
   $y_5^* = 0.0$, $y_6^* = 0.0$, $y_7^* = 0.0$, $y_8^* = 0.120001$.
5. $x_1^* = 0.0$, $x_2^* = 0.0$, $x_3^* = 1.0$, $x_4^* = 0.0$,
   $y_1^* = 0.0$, $y_2^* = 0.0$, $y_3^* = 1.0$, $y_4^* = 0.0$, $y_5^* = 0.0$, $y_6^* = 0.0$, $y_7^* = 0.0$, $y_8^* = 0.0$.

4.3 Problem 3 ($m = 5, n = 9$)

The dimension of the third problem is 16 variables.

$$A = \begin{pmatrix} 11 & 10 & 22 & 19 & 20 & 42 & 8 & 30 & 43 \\ 44 & 39 & 19 & 8 & 27 & 38 & 15 & 9 & 8 \\ 7 & 39 & 44 & 18 & 41 & 30 & 33 & 40 & 31 \\ 11 & 45 & 35 & 33 & 27 & 26 & 6 & 18 & 15 \\ 13 & 47 & 42 & 32 & 13 & 34 & 17 & 41 & 42 \end{pmatrix}, \quad B = \begin{pmatrix} 41 & 15 & 19 & 26 & 21 & 41 & 21 & 44 & 50 \\ 2 & 12 & 18 & 19 & 42 & 1 & 41 & 1 & 1 \\ 18 & 4 & 36 & 39 & 41 & 39 & 20 & 16 & 37 \\ 44 & 13 & 17 & 31 & 25 & 42 & 23 & 41 & 4 \\ 20 & 38 & 50 & 30 & 28 & 45 & 26 & 30 & 15 \end{pmatrix}.$$

When solving this problem, the implemented algorithm managed to find three anti-Berge equilibrium points (see Table 1). Each of the solutions below was obtained in two seconds of processor time.

Solutions:

1. $x_1^* = 0.0$, $x_2^* = 1.0$, $x_3^* = 0.0$, $x_4^* = 0.0$, $x_5^* = 0.0$,
   $y_1^* = 1.0$, $y_2^* = 0.0$, $y_3^* = 0.0$, $y_4^* = 0.0$,
   $y_5^* = 0.0$, $y_6^* = 0.0$, $y_7^* = 0.0$, $y_8^* = 0.0$, $y_9^* = 0.0$.
2. $x_1^* = 0.0$, $x_2^* = 0.2$, $x_3^* = 0.8$, $x_4^* = 0.0$, $x_5^* = 0.0$,
   $y_1^* = 0.0$, $y_2^* = 0.692308$, $y_3^* = 0.307692$, $y_4^* = 0.0$,
   $y_5^* = 0.0$, $y_6^* = 0.0$, $y_7^* = 0.0$, $y_8^* = 0.0$, $y_9^* = 0.0$.
3. $x_1^* = 0.0$, $x_2^* = 0.864865$, $x_3^* = 0.135135$, $x_4^* = 0.0$, $x_5^* = 0.0$,
   $y_1^* = 0.333334$, $y_2^* = 0.666666$, $y_3^* = 0.0$, $y_4^* = 0.0$,
   $y_5^* = 0.0$, $y_6^* = 0.0$, $y_7^* = 0.0$, $y_8^* = 0.0$, $y_9^* = 0.0$.

4.4 Problem 4 ($m = 6, n = 10$)

The statement of the fourth problem is more complex. The number of rows and columns of matrices A, B is increased by one compared to the previous problem.

$$A = \begin{pmatrix} 36 & 23 & 44 & 50 & 7 & 3 & 26 & 48 & 48 & 36 \\ 1 & 35 & 31 & 34 & 13 & 11 & 37 & 42 & 41 & 35 \\ 35 & 3 & 25 & 40 & 33 & 23 & 45 & 19 & 35 & 4 \\ 18 & 29 & 29 & 32 & 48 & 47 & 40 & 17 & 33 & 3 \\ 42 & 42 & 50 & 19 & 40 & 33 & 30 & 18 & 2 & 12 \\ 30 & 50 & 48 & 48 & 17 & 41 & 35 & 43 & 35 & 25 \end{pmatrix}, \quad B = \begin{pmatrix} 20 & 11 & 46 & 11 & 1 & 40 & 7 & 21 & 16 & 38 \\ 5 & 20 & 46 & 34 & 34 & 17 & 28 & 37 & 5 & 24 \\ 44 & 50 & 44 & 45 & 49 & 26 & 7 & 39 & 10 & 15 \\ 2 & 35 & 29 & 8 & 29 & 2 & 11 & 8 & 12 & 32 \\ 24 & 7 & 27 & 26 & 45 & 33 & 3 & 22 & 1 & 14 \\ 4 & 46 & 8 & 26 & 13 & 14 & 19 & 20 & 11 & 20 \end{pmatrix}.$$

When finding two of the four equilibrium points, the algorithm obtained the exact value of the quality criterion $F_* = 0.0$ (see Table 1).

Solutions:

1. $x_1^* = 0.0$, $x_2^* = 0.0$, $x_3^* = 0.0$, $x_4^* = 0.0$, $x_5^* = 0.2$, $x_6^* = 0.8$,
   $y_1^* = 0.0$, $y_2^* = 0.327586$, $y_3^* = 0.672414$, $y_4^* = 0.0$, $y_5^* = 0.0$,
   $y_6^* = 0.0$, $y_7^* = 0.0$, $y_8^* = 0.0$, $y_9^* = 0.0$, $y_{10}^* = 0.0$.
2. $x_1^* = 0.272052$, $x_2^* = 0.0$, $x_3^* = 0.0$, $x_4^* = 0.717735$, $x_5^* = 0.010213$, $x_6^* = 0.0$,
   $y_1^* = 0.0$, $y_2^* = 0.0$, $y_3^* = 0.0$, $y_4^* = 0.270811$, $y_5^* = 0.116537$,
   $y_6^* = 0.0$, $y_7^* = 0.0$, $y_8^* = 0.0$, $y_9^* = 0.612652$, $y_{10}^* = 0.0$.

3. $x_1^* = 0.0$, $x_2^* = 0.0$, $x_3^* = 0.0$, $x_4^* = 0.96$, $x_5^* = 0.0$, $x_6^* = 0.04$,
   $y_1^* = 0.0$, $y_2^* = 0.0$, $y_3^* = 0.0$, $y_4^* = 0.0$, $y_5^* = 0.428570$,
   $y_6^* = 0.571430$, $y_7^* = 0.0$, $y_8^* = 0.0$, $y_9^* = 0.0$, $y_{10}^* = 0.0$.
4. $x_1^* = 0.218440$, $x_2^* = 0.0$, $x_3^* = 0.0$, $x_4^* = 0.715347$, $x_5^* = 0.0$, $x_6^* = 0.066213$,
   $y_1^* = 0.0$, $y_2^* = 0.0$, $y_3^* = 0.0$, $y_4^* = 0.197675$, $y_5^* = 0.470930$,
   $y_6^* = 0.331395$, $y_7^* = 0.0$, $y_8^* = 0.0$, $y_9^* = 0.0$, $y_{10}^* = 0.0$.

4.5 Problem 5 ($m = 9, n = 15$)

Despite the largest dimension of this problem compared to those considered ones (26 variables), five equilibrium points were successfully obtained using the developed modification of the parallel tangent algorithm.

$$A = \begin{pmatrix} 36 & 6 & 22 & 7 & 19 & 46 & 40 & 5 & 17 & 47 & 20 & 47 & 6 & 38 & 7 \\ 5 & 27 & 24 & 44 & 22 & 2 & 6 & 34 & 34 & 36 & 10 & 26 & 11 & 21 & 3 \\ 37 & 24 & 20 & 31 & 40 & 4 & 16 & 25 & 4 & 36 & 37 & 10 & 40 & 36 & 46 \\ 32 & 2 & 38 & 17 & 16 & 41 & 38 & 49 & 6 & 48 & 25 & 11 & 19 & 39 & 42 \\ 35 & 1 & 14 & 44 & 23 & 43 & 12 & 31 & 35 & 7 & 46 & 18 & 41 & 47 & 33 \\ 44 & 49 & 9 & 42 & 33 & 19 & 18 & 21 & 33 & 29 & 5 & 18 & 48 & 33 & 19 \\ 31 & 2 & 43 & 24 & 24 & 39 & 20 & 5 & 9 & 18 & 40 & 37 & 27 & 9 & 49 \\ 42 & 44 & 15 & 22 & 39 & 14 & 23 & 15 & 42 & 26 & 5 & 4 & 38 & 2 & 39 \\ 11 & 21 & 26 & 5 & 1 & 41 & 27 & 29 & 2 & 22 & 40 & 20 & 43 & 19 & 29 \end{pmatrix}$$

$$B = \begin{pmatrix} 47 & 42 & 46 & 45 & 39 & 18 & 3 & 27 & 16 & 6 & 8 & 49 & 18 & 41 & 43 \\ 19 & 23 & 21 & 19 & 47 & 19 & 8 & 8 & 15 & 47 & 14 & 5 & 20 & 28 & 18 \\ 18 & 30 & 33 & 5 & 1 & 25 & 19 & 38 & 20 & 15 & 12 & 33 & 41 & 20 & 36 \\ 28 & 15 & 19 & 49 & 44 & 40 & 33 & 46 & 45 & 13 & 22 & 40 & 30 & 47 & 15 \\ 11 & 38 & 24 & 43 & 46 & 23 & 31 & 23 & 17 & 16 & 7 & 30 & 21 & 16 & 31 \\ 23 & 23 & 22 & 49 & 6 & 21 & 7 & 3 & 41 & 33 & 3 & 11 & 31 & 21 & 9 \\ 46 & 34 & 19 & 26 & 49 & 10 & 23 & 27 & 48 & 1 & 13 & 17 & 12 & 18 & 7 \\ 7 & 49 & 13 & 26 & 50 & 21 & 14 & 4 & 36 & 19 & 10 & 20 & 36 & 49 & 42 \\ 5 & 22 & 15 & 44 & 31 & 3 & 6 & 13 & 28 & 12 & 41 & 32 & 4 & 32 & 39 \end{pmatrix}.$$

Solutions:

1. $x_1^* = 0.0$, $x_2^* = 0.0$, $x_3^* = 0.0$, $x_4^* = 0.0$, $x_5^* = 0.0$,
   $x_6^* = 0.0$, $x_7^* = 0.0$, $x_8^* = 0.0$, $x_9^* = 1.0$,
   $y_1^* = 0.0$, $y_2^* = 0.0$, $y_3^* = 0.0$, $y_4^* = 0.0$, $y_5^* = 0.0$,
   $y_6^* = 0.0$, $y_7^* = 0.0$, $y_8^* = 0.0$, $y_9^* = 0.0$, $y_{10}^* = 0.0$,
   $y_{11}^* = 0.0$, $y_{12}^* = 0.0$, $y_{13}^* = 1.0$, $y_{14}^* = 0.0$, $y_{15}^* = 0.0$.
2. $x_1^* = 0.0$, $x_2^* = 0.407894$, $x_3^* = 0.0$, $x_4^* = 0.0$, $x_5^* = 0.0$,
   $x_6^* = 0.0$, $x_7^* = 0.136513$, $x_8^* = 0.0$, $x_9^* = 0.455593$,
   $y_1^* = 0.0$, $y_2^* = 0.0$, $y_3^* = 0.0$, $y_4^* = 0.0$, $y_5^* = 0.0$,
   $y_6^* = 0.0$, $y_7^* = 0.0$, $y_8^* = 0.228797$, $y_9^* = 0.0$, $y_{10}^* = 0.0$,
   $y_{11}^* = 0.260355$, $y_{12}^* = 0.0$, $y_{13}^* = 0.510848$, $y_{14}^* = 0.0$, $y_{15}^* = 0.0$.
3. $x_1^* = 0.0$, $x_2^* = 0.0$, $x_3^* = 0.0$, $x_4^* = 0.0$, $x_5^* = 1.0$,
   $x_6^* = 0.0$, $x_7^* = 0.0$, $x_8^* = 0.0$, $x_9^* = 0.0$,

$y_1^* = 0.0,\ y_2^* = 0.0,\ y_3^* = 0.0,\ y_4^* = 0.0,\ y_5^* = 0.0,$
$y_6^* = 0.0,\ y_7^* = 0.0,\ y_8^* = 0.0,\ y_9^* = 0.0,\ y_{10}^* = 0.0,$
$y_{11}^* = 0.0,\ y_{12}^* = 0.0,\ y_{13}^* = 0.0,\ y_{14}^* = 1.0,\ y_{15}^* = 0.0.$

4. $x_1^* = 0.0,\ x_2^* = 0.0,\ x_3^* = 0.0,\ x_4^* = 0.0,\ x_5^* = 0.0,$
$x_6^* = 0.0,\ x_7^* = 1.0,\ x_8^* = 0.0,\ x_9^* = 0.0,$
$y_1^* = 0.0,\ y_2^* = 0.0,\ y_3^* = 0.0,\ y_4^* = 0.0,\ y_5^* = 0.0,$
$y_6^* = 0.0,\ y_7^* = 0.0,\ y_8^* = 0.0,\ y_9^* = 0.0,\ y_{10}^* = 0.0,$
$y_{11}^* = 0.0,\ y_{12}^* = 0.0,\ y_{13}^* = 0.0,\ y_{14}^* = 0.0,\ y_{15}^* = 1.0.$

5. $x_1^* = 0.0,\ x_2^* = 0.0,\ x_3^* = 0.0,\ x_4^* = 0.0,\ x_5^* = 0.0,$
$x_6^* = 0.0,\ x_7^* = 0.5,\ x_8^* = 0.0,\ x_9^* = 0.5,$
$y_1^* = 0.0,\ y_2^* = 0.0,\ y_3^* = 0.0,\ y_4^* = 0.0,\ y_5^* = 0.0,$
$y_6^* = 0.8,\ y_7^* = 0.0,\ y_8^* = 0.0,\ y_9^* = 0.0,\ y_{10}^* = 0.0,$
$y_{11}^* = 0.2,\ y_{12}^* = 0.0,\ y_{13}^* = 0.0,\ y_{14}^* = 0.0,\ y_{15}^* = 0.0.$

It took three seconds of processor time to find the anti-Berge equilibrium. Therefore, we can expect that the proposed algorithm will effectively solve higher-dimensional problems.

5 Conclusion

By solving a non-local minimization problem, we have examined the two-player game for anti-Berge equilibrium. A new global search strategy based on the parallel tangent technique has been developed. It employs the one-dimensional non-local exploration procedure relying on the Strongin and parabolic techniques. The developed algorithms were implemented and investigated on several problems of different dimensions, created by the GAMUT suite. Anti-Berge equilibria were obtained correctly in all cases.

While the results for bimatrix games are promising, the applicability of the proposed algorithm to larger multiplayer games remains to be tested and represents a primary direction for future research. In this stage, we could use other approaches to building the test problems based on random generation, but we immediately focused on generally recognized serious game building tools, such as GAMUT. This is because we plan to address higher-dimensional problems and enhance the proposed approach to enable faster and more accurate searches for anti-Berge equilibrium in complex game theoretic problems.

Acknowledgments. This research was supported by the grant from the Ministry of Science and Higher Education of the Russian Federation, the project No.: 121041300060-4.

References

1. Nash, J.F.: Equilibrium points in n-person games. Proc. Nat. Acad. Sci. USA **36**, 48–49 (1950)
2. Nash, J.F.: Non-cooperative games. Ann. Math. **54**(2), 286–295 (1951)

3. Dickhaut, J., Kaplan, T.: A program for finding Nash equilibria. Mathematica J. **1**, 87–93 (1991)
4. Abalo, K.Y., Kostreva, M.M.: Some existence theorems of Nash and Berge equilibria. Appl. Math. Lett. **17**, 569–573 (2004)
5. Orlov, A.V., Strekalovsky, A.S., Batbileg, S.: On computational search for Nash equilibrium in hexamatrix games. Optim. Lett. **10**(2), 369–381 (2014). https://doi.org/10.1007/s11590-014-0833-8
6. Berge, C.: Theorie generale des jeux n-personnes. Gauthier Villars, Paris (1957)
7. Vaisman, K.S.: Berge equilibrium. Ph.D. thesis, St. Petersburg State University, St. Petersburg (1995)
8. Abalo, K.Y., Kostreva, M.M.: Berge equilibrium: some recent results from fixed-point theorems. Appl. Math. Comput. **169**, 624–638 (2005)
9. Colman, A.M., Korner, T.W., Musy, O., Tazdait, T.: Mutual support in games: some properties of Berge equilibria. J. Math. Psychol. **55**(2), 166–175 (2011)
10. Crettez, B.: On Sugden's mutually beneficial practice and Berge equilibrium. Int. Rev. Econ. **64**(4), 357–366 (2017)
11. Salukvadze, M.E., Zhukovskiy, V.I.: The Berge equilibrium: A Game-Theoretic Framework for the Golden Rule of Ethics. Birkhäuser, Cham (2020)
12. Enkhbat, R., Sukhee, B.: Optimization approach to Berge equilibrium for bimatrix game. Optim. Lett. **15**(2), 711–718 (2021). https://doi.org/10.1007/s11590-020-01688-8
13. Von Neumann, J., Morgenstern, O.: Theory of Games and Economic Behavior. Princeton University Press, Princeton (1944)
14. Tucker, A.: A two-person dilemma. In: Readings in Games and Information, pp. 7–8. Stanford University, Stanford (1950)
15. Mills, H.: Equilibrium points in finite games. J. Soc. Indust. Appl. Mathemat. **8**(2), 397–402 (1960)
16. Lemke, C.E., Howson, T.T.: Equilibrium points of bimatrix games. SIAM J. Appl. Math. **12**, 413–423 (1961)
17. Mangasarian, O.L.: Equilibrium points in bimatrix games. J. Soc. Indust. Appl. Mathemat. **12**, 778–780 (1964)
18. Mangasarian, O.L., Stone, H.: Two-person nonzero games and quadratic programming. J. Mathemat. Anal. Appl. **9**, 348–355 (1964)
19. Yanovskaya, E.B.: Equilibrium points in polymatrix games. Latvian Math. Collect. **8**, 381–384 (1968)
20. Owen, G.: Game Theory. Saunders, Philadelphia (1971)
21. Howson, J.T.: Equilibria of polymatrix games. Management Sci. **18**, 312–318 (1972)
22. Germeyer, Y.U.B.: Introduction to Operation Research. Nauka, Moscow (1976)
23. Vorobyev, N.N.: Noncooperative games. Nauka, Moscow (1984)
24. Zhukovskiy, V.I.: Some problems of non-antagonistic differential games. In: Mathematical Methods in Operation Research, pp. 103–195. Bulgarian Academy of Sciences, Sofia (1985)
25. Gibbons, R.: Game Theory for Applied Economists. Princeton University Press, Princeton (1992)
26. Zhukovskii, V.I., Chikrii, A.A.: Linear-Quadratic Differential Games, pp. 119–142. Naukova Dumka, Kiev (1994). (in Russian)
27. McKelvey, R.D., McLennan, A.: Computation of equilibria in finite games, chap. 2. In: Amman, H.M., Kendrick, D.A., Rust, J. (eds.) Handbook of Computational Economics, vol. 1, pp. 87–142. Elsevier, Amsterdam (1996)

28. Aubin, J.P.: Optima and Equilibria. Springer, Heidelberg (1998)
29. Strekalovsky, A.S.: Global optimality conditions for nonconvex optimization. J. Global Optim. **12**, 415–434 (1998)
30. Strekalovsky, A.S., Orlov, A.V.: Bimatrix Game and Bilinear Programming. Nauka, Moscow (2007)
31. Antipin, A.S.: Equilibrium programming: models and solution methods. Bull. Irkutsk State Univ. Ser. Math. **2**(1), 8–36 (2009)
32. Strekalovskii, A.S., Enkhbat, R.: Polymatrix games and optimization problems. Autom. Remote. Control. **75**(4), 632–645 (2014). https://doi.org/10.1134/S0005117914040043
33. Enkhbat, R., Tungalag, N., Gornov, A., Anikin, A.: The curvilinear search algorithm for solving three-person game. In: Proceedings of the DOOR 2016, CEUR-WS, vol. 1623, pp. 574–583 (2016)
34. Enkhbat, R., Batbileg, S., Anikin, A., Tungalag, N., Gornov, A.: A note on solving 5-person game. Adv. Model. Optim. **19**(2), 227–232 (2017)
35. Enkhbat, R., Batbileg, S., Tungalag, N., Anikin, A., Gornov, A.: A computational method for solving N-person game. Bull. Irkutsk State Univ. Ser. Math. **20**, 3–13 (2017)
36. Enkhbat, R.: A note on anti-Berge equilibrium for bimatrix game. Bull. Irkutsk State Univ. Ser. Math. **36**, 3–13 (2021)
37. Himmelblau, D.: Applied Nonlinear Programming. McGraw-Hill, New York (1972)
38. Strongin, R.: Numerical Methods in Multiextremal Problems. Nauka, Moscow (1978). (in Russian)
39. Nocedal, J., Wright, S.J.: Numerical Optimization. Springer Series in Operations Research. Springer, New York (2006)
40. Strongin, R.G., Sergeyev, Y.D.: Global Optimization with Non-Convex Constraints: Sequential and Parallel Algorithms. Kluwer Academic Publishers, Dordrecht (2000)
41. GAMUT – suite of game generators. http://gamut.stanford.edu. Accessed 28 May 2025

Embedding K Edge-Disjoint Hamiltonian Cycles on Folded Hypercubes

Kung-Jui Pai(✉)

Department of Industrial Engineering and Management, Ming Chi University of Technology, New Taipei City, Taiwan
poter@mail.mcut.edu.tw

Abstract. The n-dimensional folded hypercube FQ_n is a well-known extension of the n-dimensional hypercube that can be constructed by adding an edge to every pair of vertices with complementary addresses. The existence of the Hamiltonian cycle provides advantages for implementing algorithms that require a ring structure. In addition, $k(\geq 2)$ edge-disjoint Hamiltonian cycles also provide higher transmission efficiency for all-to-all communication algorithms and provide fault-tolerant routing for network transmission. This paper proves that there exist two edge-disjoint Hamiltonian cycles on FQ_n while $n \geq 3$; and for k edge-disjoint Hamiltonian cycles on FQ_n, the maximum k is 2 while $n \in \{3, 4\}$ and 3 while $n \in \{5, 6\}$.

Keywords: Hamiltonian Cycles · Edge-disjoint Hamiltonian Cycles · Folded Hypercubes · Interconnection Networks

1 Introduction

Interconnect topology plays an important role in achieving high performance in parallel computing systems and data centers. An interconnect topology is usually modeled by a graph in which vertices represent processing units and edges represent communication links. In the past, many interconnection topologies have been proposed, the most popular of which are torus, mesh, hypercube, hypercube variants, etc.

The hypercube [15] has become one of the most popular interconnected networks due to its recursive structure, vertex symmetry, edge symmetry, small diameter, strong connectivity, and low link complexity. The folded hypercube [4] is an extension of the hypercube constructed by adding an edge to each pair of vertices that are farthest apart, and it has been shown to improve the performance of the hypercube in many measures [4,16].

A Hamiltonian cycle in a graph is a cycle that visits every vertex exactly once. Furthermore, if $k(\geq 2)$ Hamiltonian cycles in a graph do not share any common edges, they are called edge-disjoint. The Hamiltonian decomposition of a given network is one of the important topological properties. For the case of edge-disjoint decomposition, we need to find the maximum number of edge-disjoint Hamiltonian cycles for a given network.

A. Chinchuluun et al. (Eds.): AAIM 2025, LNCS 16379, pp. 73–85, 2026.
https://doi.org/10.1007/978-981-95-5657-1_7

Many communication algorithms, such as all-to-all broadcasting algorithms, are designed based on Hamilton cycles, and their benefits can be found in [10]. Then, k edge-disjoint Hamiltonian cycles also provide some applications, which are briefly introduced below. Consider an all-to-all communication algorithm, where each vertex sends a message to all other vertices. Assume that the number of vertices in the network is m. In a single I/O port model, each vertex can send and receive one message at a time. First, a Hamiltonian cycle is formed in the network, and then at the ith step, $i \in \{1, 2, 3, \ldots, m-1\}$, each vertex can send the message received from the previous vertex in the $(i-1)^{th}$ step to the next vertex while keeping a copy of the message. This all-to-all communication algorithm requires $m-1$ steps. Now, assume that each vertex has d I/O ports, which means that each vertex can send and receive d messages per unit time. In this case, if we can construct $k(\leq d/2)$ edge-disjoint Hamiltonian cycles, then each message of size q can be split into messages of size q/k, and each message can be sent along the k edge-disjoint Hamiltonian cycles. Therefore, the time complexity of the algorithm can be improved by a factor of k.

Another application of k edge-disjoint Hamiltonian cycles is useful in the context of fault-tolerant routing. If there exist k edge-disjoint Hamiltonian cycles in the network, then a vertex can still communicate with any other vertex even in the presence of $2k-1$ edge failures.

Edge-disjoint Hamiltonian cycles have been studied in various networks for nearly two decades. The networks that have been studied are: de Bruijn graphs [14], Eisenstein-Jacobi networks [9], Gaussian Networks [1], butterfly networks [3], k-ary n-cubes [2], hypercubes [2], hypertournaments [13], locally twisted cubes [5,11], crossed cubes [7,12], augmented cubes [6], twisted cubes [8], parity cubes [17], transposition networks [7], and hypercube-like networks [7]. To the best of my knowledge, this paper is the first to address this problem on the n-dimensional folded hypercube (FQ_n). Similar to locally twisted cubes, crossed cubes, and augmented cubes, FQ_n is a well-studied variant of the standard hypercube. These variants substantially reduce the diameter (communication delay) and enhance connectivity and fault tolerance compared to the conventional hypercube. In the same n-dimensional topology, the degree of FQ_n is $n+1$, whereas that of locally twisted and crossed cubes is n. This additional edge per node provides FQ_n with sufficient connectivity to construct edge-disjoint Hamiltonian cycles. For example, Fig. 2 shows that FQ_3 contains two edge-disjoint Hamiltonian cycles.

The rest of the paper is organized as follows: In Sect. 2, the structure of folded hypercubes is introduced, and some notations are given. Section 3 shows how to construct two edge-disjoint Hamiltonian cycles in FQ_n while $n \geq 3$, and proves there exist three edge-disjoint Hamiltonian cycles on FQ_5 and FQ_6. Finally, Sect. 4 is the concluding remarks of this paper.

2 Preliminaries

Interconnection networks are usually modeled as undirected simple graphs $G = (V, E)$, where the vertex set $V(= V(G))$ and the edge set $E(= E(G))$ represent the set of processing units and the set of communication links between vertices, respectively. A path P_k of length k in G, denoted by v_0 - v_1 - v_2 - ... - v_{k-2} - v_{k-1} - v_k, is a sequence $(v_0, v_1, v_2, \ldots, v_{k-1}, v_k)$ of vertices such that $(v_i, v_{i+1}) \in E$ for $0 \leq i \leq k-1$. A cycle C_k of length k in G, denoted by v_0 - v_1 - v_2 - ... - v_{k-2} - v_{k-1} - v_0, is a sequence $(v_0, v_1, v_2, \ldots, v_{k-1}, v_0)$ of vertices such that $(v_{k-1}, v_0) \in E$ and $(v_i, v_{i+1}) \in E$ for $0 \leq i \leq k-2$. Now, we introduce hypercubes and folded hypercubes as follows:

Definition 1 *(Saad and Schultz [15]). The n-dimensional hypercube, denoted by Q_n, is a graph with 2^n vertices such that each vertex corresponds to an n-tuple $(b_{n-1}, b_{n-2}, \ldots, b_1, b_0)$, on the set $\{0, 1\}^n$ and two vertices are linked by an edge if and only if they differ in exactly one coordinate.*

Definition 2 *(El-Amawy and Latifi [4]). The n-dimensional folded hypercube, denoted by FQ_n, is a graph with 2^n vertices such that each vertex corresponds to an n-tuple $(b_{n-1}, b_{n-2}, \ldots, b_1, b_0)$ on the set $\{0, 1\}^n$ and two vertices are linked by an edge if and only if they differ in exactly one coordinate or a complementary edges if and only if they differ in each coordinate.*

For convenience, this article will call the edge connecting two vertices whose coordinates differ by exactly one as a hypercube edge. For example, Fig. 1(a) and (b) show the 3-dimensional hypercube Q_3 and the 3-dimensional folded hypercube FQ_3, respectively.

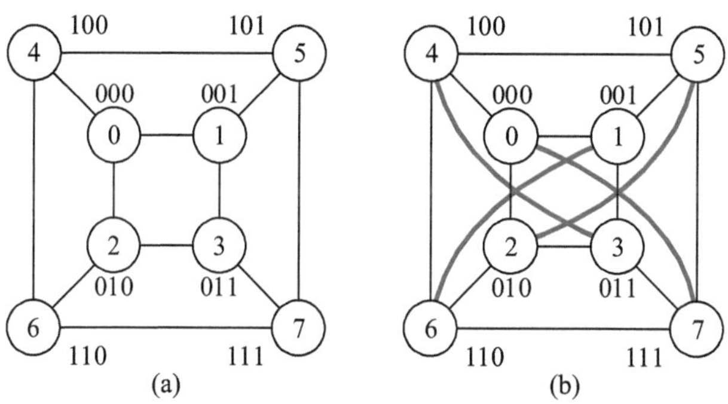

Fig. 1. (a) The 3-dimensional hypercube Q_3 and (b) the 3-dimensional folded hypercube FQ_3, where the thick red lines indicate complementary edges and vertex labels are expressed in both binary and decimal.

3 Main Results

First, we find that FQ_3 has only 8 vertices and 16 edges, and these edges easily split into 2 edge-disjoint Hamiltonian cycles. Then, we get the following lemma.

Lemma 1. *There exist two edge-disjoint Hamiltonian cycles on* FQ_3

Proof. Let $C_8^A = 2 - 0 - 1 - 3 - 4 - 6 - 7 - 5 - 2$ and $C_8^B = 0 - 4 - 5 - 1 - 6 - 2 - 3 - 7 - 0$. Both C_8^A and C_8^B are cycles that visited each vertex exactly once, and their edge-disjointness can be visually observed in Fig. 2(a). Therefore, C_8^A and C_8^B form two edge-disjoint Hamiltonian cycles in FQ_3. □

By Definitions 1 and 2, FQ_4 consists of two Q_3s, 8 new hypercube edges and 8 complementary edges. According to the order of C_8^A visited vertices, when starting from vertices 2 and 10 respectively, we can get two C_8s on FQ_4. After adding and deleting two edges from two C_8s, we can get a C_{16} on FQ_4. In the same way, using the order of C_8^B visited vertices, we can get another C_{16} on FQ_4. In addition, we found that two complementary edges are used in each of these two cycles C_8^A and C_8^B. The method is formally written as the following algorithm.

Algorithm HCycleForNextFQ

Input: n, HC_A // n: the dimensional of FQ_n, HC_A: a Hamiltonian cycle with two complementary edges in FQ_n

Output: HC_B // a Hamiltonian cycle with two complementary edges in FQ_{n+1}

1. $m = 2^n$; // number of vertices in FQ_n
2. Let p_1 and p_2 be the paths obtained by removing two complementary edges of HC_A, where p_1 is represented as $v_1 - v_2 - v_3 - \ldots - v_k$ and p_2 is represented as $v_{k+1} - v_{k+2} - v_{k+3} - \ldots - v_m$.
3. $p_3 = (v_1 + 2^n) - (v_2 + 2^n) - (v_3 + 2^n) - \ldots - (v_k + 2^n)$;
4. $p_4 = (v_{k+1} + 2^n) - (v_{k+2} + 2^n) - (v_{k+3} + 2^n) - \ldots - (v_m + 2^n)$;
5. $rp_2 =$ reversed(p_2) ; // i.e. $rp_2 = v_m - v_{m-1} - v_{m-2} - \ldots - v_{k+2} - v_{k+1}$
6. $rp_3 =$ reversed(p_3) ; // i.e. $rp_3 = (v_k + 2^n) - (v_{k-1} + 2^n) - \ldots - (v_1 + 2^n)$
7. $HC_B = p_1 + rp_3 + rp_2 + p_4 + v_1$; // + for concatenate
8. **return** HC_B;

Lemma 2. *The vertex sequence generated by Algorithm HCycleForNextFQ forms a Hamiltonian cycle with two complementary edges in* FQ_{n+1}.

Proof. According to Definitions 1 and 2, FQ_{n+1} consists of two Q_ns, called Q_n^0 and Q_n^1 respectively. The input of the algorithm is a Hamiltonian cycle HC_A with two complementary edges in FQ_n. Removing two complementary edges of HC_A to obtain p_1 and p_2. Then the vertex union of p_1 and p_2 is the vertex set of Q_n^0. p_3 (p_4, respectively) is obtained by adding 2^n to each vertex label in the vertex sequence of p_1 (p_2, respectively). So the vertex union of p_3 and p_4 is the vertex set of Q_n^1. Therefore, p_1, p_2, p_3, and p_4 visit each vertex of FQ_{n+1} exactly once.

According Algorithm HCycleForNextFQ, $HC_B = p_1 + rp_3 + rp_2 + p_4 + v_1$. There exists a hypercube edge $(v_k, v_k + 2^n)$ between the last vertex v_k of p_1 and the first vertex $(v_k + 2^n)$ of reversal p_3. Next, $(v_1 + 2^n, v_m)$ connects the last vertex $(v_1 + 2^n)$ of reversal p_3 and the first vertex v_m of reversed p_2. Because (v_1, v_m) is a complementary edge in FQ_n, $(v_1 + 2^n, v_m)$ is a complementary edge in FQ_{n+1}. For similar reasons, $(v_{k+1}, v_{k+1} + 2^n)$ is a hypercube edge and $(v_m + 2^n, v_1)$ is a complementary edge. With these 4 edges, we can connect the 4 paths p_1, p_3, p_2 and p_4 into a cycle HC_B. These 4 edges visit each vertex of FQ_{n+1} exactly once, so HC_B is a Hamiltonian cycle with two complementary edges in FQ_{n+1}. □

For example, input $n = 3$ and $C_8^A = 2 - 0 - 1 - 3 - 4 - 6 - 7 - 5 - 2$ into Algorithm HCycleForNextFQ. Since two complementary edges are $(3, 4)$ and $(5, 2)$, we will have $p_1 = 2 - 0 - 1 - 3$ and $p_2 = 4 - 6 - 7 - 5$. Next, $p_3 = 10 - 8 - 9 - 11$ and $p_4 = 12 - 14 - 15 - 13$. So $HC_B = 2 - 0 - 1 - 3 - 11 - 9 - 8 - 10 - 5 - 7 - 6 - 4 - 12 - 14 - 15 - 13 - 2$. The edges connecting the four paths are $(3, 11)$, $(10, 5)$, $(4, 12)$ and $(13, 2)$, respectively. It is easy to verify that (3,11) and (4, 12) are hypercube edges, and $(10, 5)$ and $(13, 2)$ are complementary edges. Therefore, $HC_B = 2 - 0 - 1 - 3 - 11 - 9 - 8 - 10 - 5 - 7 - 6 - 4 - 12 - 14 - 15 - 13 - 2$ is a Hamiltonian cycle with two complementary edges in FQ_4.

Lemma 3. *There exist two edge-disjoint Hamiltonian cycles on FQ_4.*

Proof. Substituting $n = 3$ and $C_8^A = 2 - 0 - 1 - 3 - 4 - 6 - 7 - 5 - 2$ ($C_8^B = 0 - 4 - 5 - 1 - 6 - 2 - 3 - 7 - 0$, respectively) into Algorithm HCycleForNextFQ, we obtain $C_{16}^A = 2 - 0 - 1 - 3 - 11 - 9 - 8 - 10 - 5 - 7 - 6 - 4 - 12 - 14 - 15 - 13 - 2$ ($C_{16}^B = 0 - 4 - 5 - 1 - 9 - 13 - 12 - 8 - 7 - 3 - 2 - 6 - 14 - 10 - 11 - 15 - 0$, respectively). Both C_{16}^A and C_{16}^B are Hamiltonian cycles on FQ_4. Next, the input cycles C_8^A and C_8^B are edge-disjoint, so the output cycles C_{16}^A and C_{16}^B are edge-disjoint. This lemma holds. □

For example, Fig. 2(b) shows two edge-disjoint Hamiltonian cycles C_{16}^A and C_{16}^B on FQ_4.

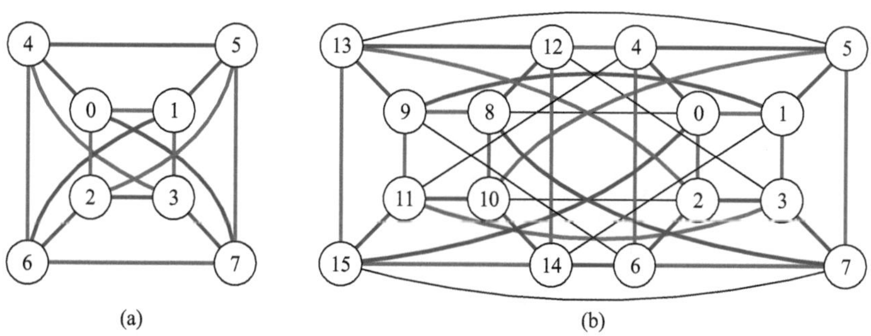

Fig. 2. (a) Two edge-disjoint Hamiltonian cycles on FQ_3 and (b) two edge-disjoint Hamiltonian cycles on FQ_4, where the thick red lines and the thick blue lines represent two edge-disjoint Hamiltonian cycles, respectively.

Theorem 1. *There exist two edge-disjoint Hamiltonian cycles on FQ_n while $n \geq 3$.*

Proof. In fact, Algorithm HCycleForNextFQ is a recursive construction method. Taking two edge-disjoint Hamiltonian cycles with two complementary edges in FQ_n as input, we can get two edge-disjoint Hamiltonian cycles with two complementary edges in FQ_{n+1}. Based on C_8^A = 2 - 0 - 1 - 3 - 4 - 6 - 7 - 5 - 2 and C_8^B = 0 - 4 - 5 - 1 - 6 - 2 - 3 - 7 - 0, two edge-disjoint Hamiltonian cycles in each dimension can be formed in sequence. □

Lemma 4. *For k edge-disjoint Hamiltonian cycles on FQ_3 and FQ_4, the maximum k is 2.*

Proof. According to Lemmas 1 and 3, the case $k \geq 2$ holds. However, the degree of FQ_3 (FQ_4, respectively) is 4 (5, respectively), after being used by two edge-disjoint Hamiltonian cycles, the number of remaining edges is 0 (8, respectively). There are not enough edges to form another Hamiltonian cycle. Therefore, the case $k \leq 2$ holds. □

Next, we try to find three edge-disjoint Hamiltonian cycles on FQ_5. According to Algorithm HCycleForNextFQ, Substituting $n = 4$ and C_{16}^A (C_{16}^B, respectively) can get C_{32}^A = 2 - 0 - 1 - 3 - 11 - 9 - 8 - 10 - 26 - 24 - 25 - 27 - 19 - 17 - 16 - 18 - 13 - 15 - 14 - 12 - 4 - 6 - 7 - 5 - 21 - 23 - 22 - 20 - 28 - 30 - 31 - 29 - 2 (C_{32}^B = 0 - 4 - 5 - 1 - 9 - 13 - 12 - 8 - 24 - 28 - 29 - 25 - 17 - 21 - 20 - 16 - 15 - 11 - 10 - 14 - 6 - 2 - 3 - 7 - 23 - 19 - 18 - 22 - 30 - 26 - 27 - 31 - 0, respectively).

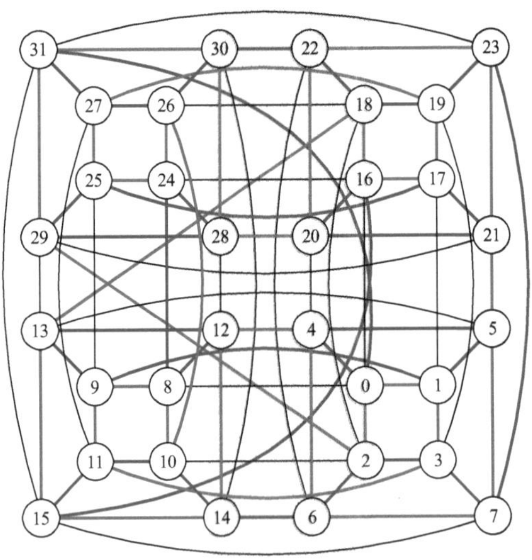

Fig. 3. The thick red and blue lines represent two edge-disjoint Hamiltonian cycles on FQ_5, respectively. To make the graph easier to read, unused complementary edges are omitted.

Figure 3 illustrates two edge-disjoint Hamiltonian cycles C_{32}^A and C_{32}^B on FQ_5. When FQ_5 removes all edges of C_{32}^A, the remaining graph is a 4-regular graph as shown in Fig. 4(a). It can still be decomposed into two edge-disjoint Hamiltonian cycles. But we adopt C_{32}^B as the second edge-disjoint Hamiltonian cycle, the remaining edges in FQ_5 will form two C_8 and four C_4. Let them be $C^1 = 0 - 8 - 23 - 31 - 15 - 7 - 24 - 16 - 0$, $C^2 = 1 - 17 - 14 - 30 - 1$, $C^3 = 2 - 10 - 21 - 29 - 13 - 5 - 26 - 18 - 2$, $C^4 = 3 - 19 - 12 - 28 - 3$, $C^5 = 4 - 20 - 11 - 27 - 4$, $C^6 = 6 - 22 - 9 - 25 - 6$. Next, the idea is to remove some edges from these cycles to produce multiple paths, and then connect these paths to form a Hamiltonian cycle. The edges used to connect the paths are taken from C_{32}^B. Therefore, the removed edges of these cycles are given to C_{32}^B to form a new Hamiltonian cycle. The third Hamiltonian cycle C_{32}^C can be constructed as: $E(C_{32}^C) = (E(C^1) \cup E(C^2) \cup E(C^3) \cup E(C^4) \cup E(C^5) \cup E(C^6) \setminus \{(0,8),(1,30),(4,27),(9,22),(9,25),(12,19),(13,29),(18,26)\}) \cup \{(0,4),(1,9),(8,12),(9,13),(18,19),(22,30),(25,29),(26,27)\}$. Final, the new second Hamiltonian cycle $C_{32}^D = (C_{32}^B \setminus \{(0,4),(1,9),(8,12),(9,13),(18,19),(22,30),(25,29),(26,27)\}) \cup \{(0,8),(1,30),(4,27),(9,22),(9,25),(12,19),(13,29),(18,26)\}$. By adding and deleting edges, Fig. 4(a) illustrates how to construct C_{32}^C from $C^1, C^2, C^3, C^4, C^5, C^6$, and Fig. 4(b) illustrates how to construct C_{32}^D from C_{32}^B. Finally, we provide Table 1 to show how FQ_5 is decomposed into three C_{32}.

Table 1. Vertex adjacency list of FQ_5, where the subscripts A, C, D represent whether the adjacent vertex is located in the cycle C_{32}^A, C_{32}^C, or C_{32}^D, respectively.

u	Adjacent vertices of u	u	Adjacent vertices of u
0	1_A, 2_A, 4_C, 8_D, 16_C, 31_D	16	0_C, 15_D, 17_A, 18_A, 20_D, 24_C
1	0_A, 3_A, 5_D, 9_C, 17_C, 30_D	17	1_C, 14_C, 16_A, 19_A, 21_D, 25_D
2	0_A, 3_D, 6_D, 10_C, 18_C, 29_A	18	2_C, 13_A, 16_A, 19_C, 22_D, 26_D
3	1_A, 2_D, 7_D, 11_A, 19_C, 28_C	19	3_C, 12_D, 17_A, 18_C, 23_D, 27_A
4	0_C, 5_D, 6_A, 12_A, 20_C, 27_D	20	4_C, 11_C, 16_D, 21_D, 22_A, 28_A
5	1_D, 4_D, 7_A, 13_C, 21_A, 26_C	21	5_A, 10_C, 17_D, 20_D, 23_A, 29_C
6	2_D, 4_A, 7_A, 14_D, 22_C, 25_C	22	6_C, 9_D, 18_D, 20_A, 23_A, 30_C
7	3_D, 5_A, 6_A, 15_C, 23_D, 24_C	23	7_D, 8_C, 19_D, 21_A, 22_A, 31_C
8	0_D, 9_A, 10_A, 12_C, 23_C, 24_D	24	7_C, 8_D, 16_C, 25_A, 26_A, 28_D
9	1_C, 8_A, 11_A, 13_C, 22_D, 25_D	25	6_C, 9_D, 17_D, 24_A, 27_A, 29_C
10	2_C, 8_A, 11_D, 14_D, 21_C, 26_A	26	5_C, 10_A, 18_D, 24_A, 27_C, 30_D
11	3_A, 9_A, 10_D, 15_D, 20_C, 27_C	27	4_D, 11_C, 19_A, 25_A, 26_C, 31_D
12	4_A, 8_C, 13_D, 14_A, 19_D, 28_C	28	3_C, 12_C, 20_A, 24_D, 29_D, 30_A
13	5_C, 9_C, 12_D, 15_A, 18_A, 29_D	29	2_A, 13_D, 21_C, 25_C, 28_D, 31_A
14	6_D, 10_D, 12_A, 15_A, 17_C, 30_C	30	1_D, 14_C, 22_C, 26_D, 28_A, 31_A
15	7_C, 11_D, 13_A, 14_A, 16_D, 31_C	31	0_D, 15_C, 23_C, 27_D, 29_A, 30_A

Lemma 5. *There exist three edge-disjoint Hamiltonian cycles on FQ_5.*

Proof. According to the above modification operation, C_{32}^A = 0 - 1 - 3 - 19 - 17 - 16 - 18 - 26 - 24 - 25 - 27 - 4 - 6 - 7 - 5 - 13 - 15 - 14 - 12 - 28 - 30 - 31 - 29 - 21 - 23 - 22 - 20 - 11 - 9 - 8 - 10 - 2 - 0, C_{32}^C = 0 - 4 - 20 - 11 - 27 - 26 - 5 - 13 - 9 - 1 - 17 - 14 - 30 - 22 - 6 - 25 - 29 - 21 - 10 - 2 - 18 - 19 - 3 - 28 - 12 - 8 - 23 - 31 - 15 - 7 - 24 - 16 - 0, and C_{32}^D = 0 - 31 - 27 - 4 - 5 - 1 - 30 - 26 - 18 - 22 - 9 - 25 - 17 - 21 - 20 - 16 - 15 - 11 - 10 - 14 - 6 - 2 - 3 - 7 - 23 - 19 - 12 - 13 - 29 - 28 - 24 - 8 - 0. All C_{32}^A, C_{32}^C and C_{32}^D are cycles that visited all vertex on FQ_5 exactly once. Since their edge assignments are based on edge-disjoint decomposition, the edge-disjoint property holds. □

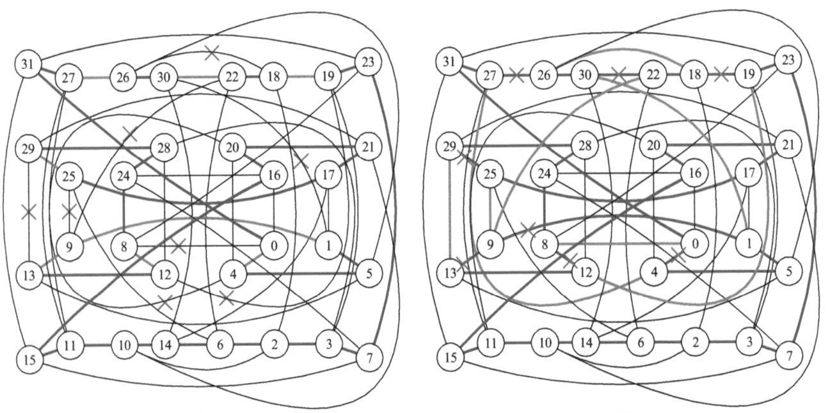

(a) (b)

Fig. 4. (a) The thin black lines represent C^1, C^2, C^3, C^4, C^5 and C^6. C_{32}^C consists of thick green lines and uncrossed black edges; (b) C_{32}^D is formed by thick green lines and uncrossed blue edges. To make the graph easier to read, all edges of C_{32}^A are omitted.

Again, Substituting $n = 5$ and C_{32}^A (C_{32}^B, respectively) in Algorithm HCycle-ForNextFQ to get C_{64}^A = 2 - 0 - 1 - 3 - 11 - 9 - 8 - 10 - 26 - 24 - 25 - 27 - 19 - 17 - 16 - 18 - 50 - 48 - 49 - 51 - 59 - 57 - 56 - 58 - 42 - 40 - 41 - 43 - 35 - 33 - 32 - 34 - 29 - 31 - 30 - 28 - 20 - 22 - 23 - 21 - 5 - 7 - 6 - 4 - 12 - 14 - 15 - 13 - 45 - 47 - 46 - 44 - 36 - 38 - 39 - 37 - 53 - 55 - 54 - 52 - 60 - 62 - 63 - 61 - 2 (C_{64}^B = 0 - 4 - 5 - 1 - 9 - 13 - 12 - 8 - 24 - 28 - 29 - 25 - 17 - 21 - 20 - 16 - 48 - 52 - 53 - 49 - 57 - 61 - 60 - 56 - 40 - 44 - 45 - 41 - 33 - 37 - 36 - 32 - 31 - 27 - 26 - 30 - 22 - 18 - 19 - 23 - 7 - 3 - 2 - 6 - 14 - 10 - 11 - 15 - 47 - 43 - 42 - 46 - 38 - 34 - 35 - 39 - 55 - 51 - 50 - 54 - 62 - 58 - 59 - 63 - 0, respectively). Figure 5 illustrates two edge-disjoint Hamiltonian cycles C_{64}^A and C_{64}^B on FQ_6.

When FQ_6 removes all edges of C_{64}^A, the remaining graph is a 5-regular graph. It can still be decomposed into two edge-disjoint Hamiltonian cycles. But we adopt C_{64}^B as the second edge-disjoint Hamiltonian cycle, the unused edges in FQ_6 will form six components H^1, H^2, H^3, H^4, H^5, H^6. Each of them is

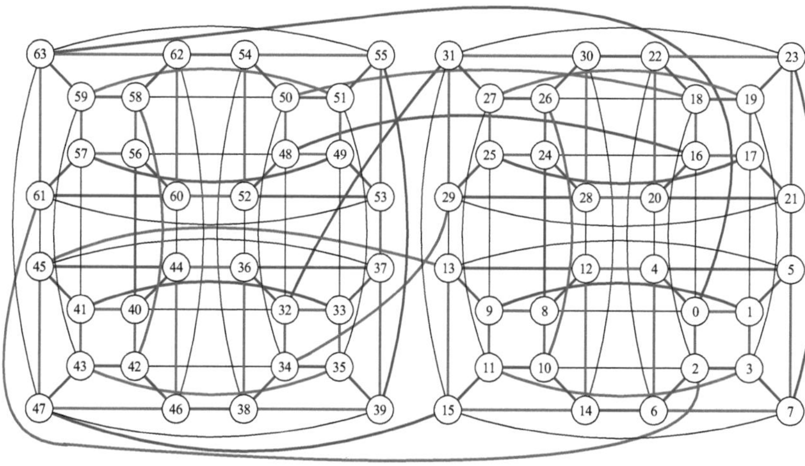

Fig. 5. The thick red and blue lines represent two edge-disjoint Hamiltonian cycles C_{64}^A and C_{64}^B on FQ_6, respectively. To make the graph easier to read, unused complementary edges and some unused hypercube edges are omitted.

a 3-regular graph. $V(H^1) = \{0, 7, 8, 15, 16, 23, 24, 31, 32, 39, 40, 47, 48, 55, 56, 63\}$, $V(H^2) = \{1, 14, 17, 30, 33, 46, 49, 62\}$, $V(H^3) = \{2, 5, 10, 13, 18, 21, 26, 29, 34, 37, 42, 45, 50, 53, 58, 61\}$, $V(H^4) = \{3, 12, 19, 28, 35, 44, 51, 60\}$, $V(H^5) = \{4, 11, 20, 27, 36, 43, 52, 59\}$, and $V(H^6) = \{6, 9, 22, 25, 38, 41, 54, 57\}$. Their edge sets are shown in Fig. 6. Since these six components are disconnected, we need to take eight edges in C_{64}^B to connect them. These eight edges are (0, 8), (1, 30), (4, 27), (9, 22), (9, 25), (12, 19), (13, 29), (18, 26). Finally, we have C_{64}^C = 0 - 8 - 40 - 23 - 55 - 63 - 31 - 15 - 11 - 43 - 20 - 52 - 36 - 4 - 59 - 27 - 26 - 18 - 2 - 10 - 53 - 61 - 45 - 37 - 5 - 58 - 62 - 1 - 17 - 46 - 14 - 49 - 33 - 30 - 22 - 6 - 38 - 25 - 57 - 41 - 9 - 54 - 50 - 13 - 29 - 21 - 42 - 34 - 35 - 51 - 12 - 28 - 60 - 44 - 19 - 3 - 7 - 56 - 24 - 39 - 47 - 16 - 48 - 32 - 0, as shown by the thick lines in Fig. 6.

Next, eight edges in C_{64}^B are taken, so we should take some unused edges in H^1, H^2, H^3, H^4, H^5, and H^6 to connect the remaining edges to form a Hamiltonian cycle. The new second Hamiltonian cycle $C_{64}^D = (C_{64}^B \setminus \{(3, 7), (11, 15), (16, 48), (22, 30), (24, 28), (26, 27), (34, 35), (35, 39), (50, 54), (58, 62)\}) \cup \{(3, 35), (7, 39), (11, 27), (15, 48), (16, 24), (22, 54), (26, 58), (28, 35), (30, 62), (34, 50)\}$, as shown in Fig. 7. Finally, we provide Table 2 to show how FQ_6 is decomposed into three C_{32}.

Lemma 6. *There exist three edge-disjoint Hamiltonian cycles on* FQ_6.

Proof. All C_{64}^A, C_{64}^C and C_{64}^D are cycles that visited all vertex on FQ_6 exactly once. Since their edge assignments are based on edge-disjoint decomposition, the edge-disjoint property holds. □

Lemma 7. *For k edge-disjoint Hamiltonian cycles on* FQ_5 *and* FQ_6*, the maximum k is 3.*

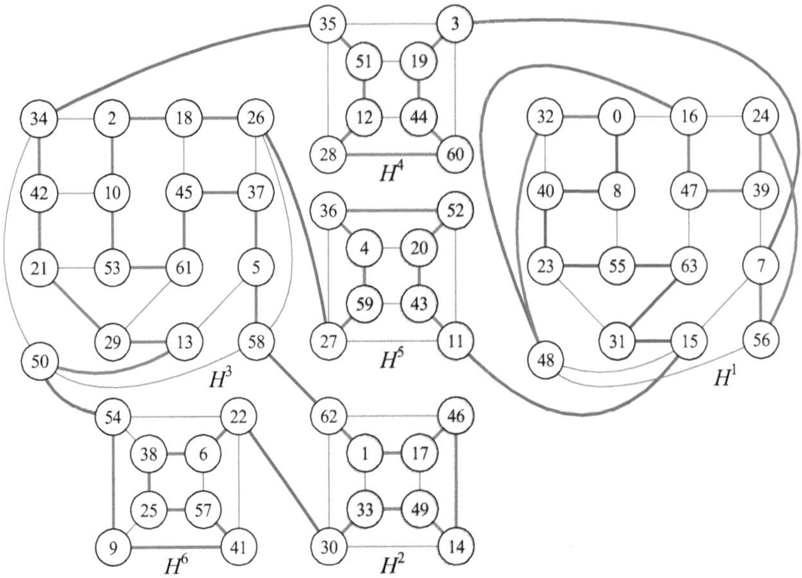

Fig. 6. Thick and thin green lines show six components H^1, H^2, H^3, H^4, H^5, H^6, and thick green and red lines represent C_{64}^C on FQ_6.

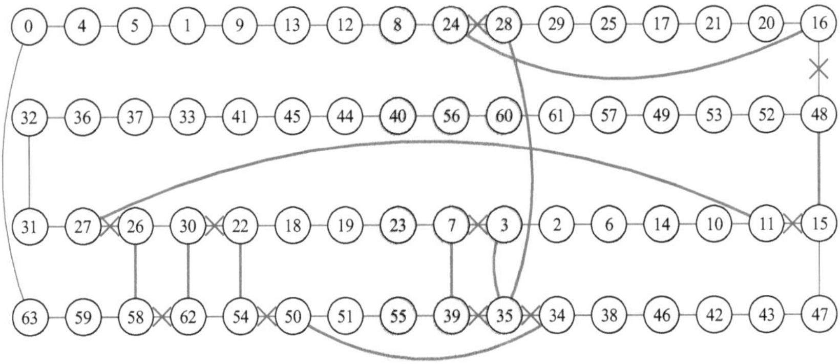

Fig. 7. The thin red lines with or without cross marks represent C_{64}^B on FQ_6, and thick green and un-marked red edges represent C_{64}^D on FQ_6.

Proof. According to Lemmas 5 and 6, the case $k \geq 3$ holds. However, the degree of FQ_5 (FQ_6, respectively) is 6 (7, respectively), and there are no (32, respectively) edges that do not serve as Hamiltonian cycles. There are not enough edges to form another Hamiltonian cycle. Therefore, the case $k \leq 3$ holds. □

Summarizing Theorem 1 and Lemmas 4 and 7, we give the following theorem at the end of this section.

Theorem 2. *FQ_n with $n \geq 3$ contains two edge-disjoint Hamiltonian cycles, and for k edge-disjoint Hamiltonian cycles on FQ_n, the maximum k is 2 while $n \in \{3,4\}$ and 3 while $n \in \{5,6\}$.*

Table 2. Vertex adjacency list of FQ_6, where the subscripts A, C, D represent whether the adjacent vertex is located in the cycle C_{64}^A, C_{64}^C, or C_{64}^D, respectively.

u	Adjacent vertices of u	u	Adjacent vertices of u
0	$1_A, 2_A, 4_D, 8_C, 16, 32_C, 63_D$	32	$0_C, 31_D, 33_A, 34_A, 36_D, 40, 48_C$
1	$0_A, 3_A, 5_D, 9_D, 17_C, 33, 62_C$	33	$1, 30_C, 32_A, 35_A, 37_D, 41_D, 49_C$
2	$0_A, 3_D, 6_D, 10_C, 18_C, 34, 61_A$	34	$2, 29_A, 32_A, 35_C, 38_D, 42_C, 50_D$
3	$1_A, 2_D, 7_C, 11_A, 19_C, 35_D, 60$	35	$3_D, 28_D, 33_A, 34_C, 39, 43_A, 51_C$
4	$0_D, 5_D, 6_A, 12_A, 20, 36_C, 59_C$	36	$4_C, 27, 32_D, 37_D, 38_A, 44_A, 52_C$
5	$1_D, 4_D, 7_A, 13, 21_A, 37_C, 58_C$	37	$5_C, 26, 33_D, 36_D, 39_A, 45_C, 53_A$
6	$2_D, 4_A, 7_A, 14_D, 22_C, 38_C, 57$	38	$6_C, 25_C, 34_D, 36_A, 39_A, 46_D, 54$
7	$3_C, 5_A, 6_A, 15, 23_D, 39_D, 56_C$	39	$7_D, 24_C, 35, 37_A, 38_A, 47_C, 55_D$
8	$0_C, 9_A, 10_A, 12_D, 24_D, 40_C, 55$	40	$8_C, 23_C, 32, 41_A, 42_A, 44_D, 56_D$
9	$1_D, 8_A, 11_A, 13_D, 25, 41_C, 54_C$	41	$9_C, 22, 33_D, 40_A, 43_A, 45_D, 57_C$
10	$2_C, 8_A, 11_D, 14_D, 26_A, 42, 53_C$	42	$10, 21_C, 34_C, 40_A, 43_D, 46_D, 58_A$
11	$3_A, 9_A, 10_D, 15_C, 27_D, 43_C, 52$	43	$11_C, 20_C, 35_A, 41_A, 42_D, 47_D, 59$
12	$4_A, 8_D, 13_D, 14_A, 28_C, 44, 51_C$	44	$12, 19_C, 36_A, 40_D, 45_D, 46_A, 60_C$
13	$5, 9_D, 12_D, 15_A, 29_C, 45_A, 50_C$	45	$13_A, 18, 37_C, 41_D, 44_D, 47_A, 61_C$
14	$6_D, 10_D, 12_A, 15_A, 30, 46_C, 49_C$	46	$14_C, 17_C, 38_D, 42_D, 44_A, 47_A, 62$
15	$7, 11_C, 13_A, 14_A, 31_C, 47_D, 48_D$	47	$15_D, 16_C, 39_C, 43_D, 45_A, 46_A, 63$
16	$0, 17_A, 18_A, 20_D, 24_D, 47_C, 48_C$	48	$15_D, 16_C, 32_C, 49_A, 50_A, 52_D, 56$
17	$1_C, 16_A, 19_A, 21_D, 25_D, 46_C, 49$	49	$14_C, 17, 33_C, 48_A, 51_A, 53_D, 57_D$
18	$2_C, 16_A, 19_D, 22_D, 26_C, 45, 50_A$	50	$13_C, 18_A, 34_D, 48_A, 51_D, 54_C, 58$
19	$3_C, 17_A, 18_D, 23_D, 27_A, 44_C, 51$	51	$12_C, 19, 35_C, 49_A, 50_D, 55_D, 59_A$
20	$4, 16_D, 21_D, 22_A, 28_A, 43_C, 52_C$	52	$11, 20_C, 36_C, 48_D, 53_D, 54_A, 60_A$
21	$5_A, 17_D, 20_D, 23_A, 29_C, 42_C, 53$	53	$10_C, 21, 37_A, 49_D, 52_D, 55_A, 61_C$
22	$6_C, 18_D, 20_A, 23_A, 30_C, 41, 54_D$	54	$9_C, 22_D, 38, 50_C, 52_A, 55_A, 62_D$
23	$7_D, 19_D, 21_A, 22_A, 31, 40_C, 55_C$	55	$8, 23_C, 39_D, 51_D, 53_A, 54_A, 63_C$
24	$8_D, 16_D, 25_A, 26_A, 28, 39_C, 56_C$	56	$7_C, 24_C, 40_D, 48, 57_A, 58_A, 60_D$
25	$9, 17_D, 24_A, 27_A, 29_D, 38_C, 57_C$	57	$6, 25_C, 41_C, 49_D, 56_A, 59_A, 61_D$
26	$10_A, 18_C, 24_A, 27_C, 30_D, 37, 58_D$	58	$5_C, 26_D, 42_A, 50, 56_A, 59_D, 62_C$
27	$11_D, 19_A, 25_A, 26_C, 31_D, 36, 59_C$	59	$4_C, 27_C, 43, 51_A, 57_A, 58_D, 63_D$
28	$12_C, 20_A, 24, 29_D, 30_A, 35_D, 60_C$	60	$3, 28_C, 44_C, 52_A, 56_D, 61_D, 62_A$
29	$13_C, 21_C, 25_D, 28_D, 31_A, 34_A, 61$	61	$2_A, 29, 45_C, 53_C, 57_D, 60_D, 63_A$
30	$14, 22_C, 26_D, 28_A, 31_A, 33_C, 62_D$	62	$1_C, 30_D, 46, 54_D, 58_C, 60_A, 63_A$
31	$15_C, 23, 27_D, 29_A, 30_A, 32_D, 63_C$	63	$0_D, 31_C, 47, 55_C, 59_D, 61_A, 62_A$

4 Concluding Remarks

The main contribution of this paper lies in the final theorem. Hypercube-like networks—such as locally twisted cubes, crossed cubes, twisted cubes, and parity cubes—are all constructed recursively. Consequently, prior research has leveraged this recursive property to construct edge-disjoint Hamiltonian cycles. In contrast, the folded hypercube FQ_n differs from these graph classes by incorporating complementary edges into the hypercube structure. However, during the recursive construction of FQ_{n+1}, the original complementary edges are removed and new ones are introduced. This characteristic complicates the construction of edge-disjoint Hamiltonian cycles in the next-dimensional folded hypercube. To address this challenge, this paper proposes a specific algorithm to construct two edge-disjoint Hamiltonian cycles on FQ_n. Next, we employ the method of exchanging some edges to construct three edge-disjoint Hamiltonian cycles.

For future work, it is interesting to examine whether there are four edge-disjoint Hamiltonian cycles in FQ_n for $n \geq 7$. Specifically, based on the structural properties of FQ_n, we conjecture that FQ_n is likely to contain at least four edge-disjoint Hamiltonian cycles for sufficiently large n, since the increased degree may support multiple edge-disjoint Hamiltonian cycles constructions. However, a formal proof of this conjecture remains an open problem and is left for future work.

Acknowledgments. This work is supported by NSTC 113-2221-E-131-035 from the National Science and Technology Council, Taiwan.

References

1. Albader, B., Bose, B.: Edge disjoint Hamiltonian cycles in Gaussian networks. IEEE Trans. Comput. **65**, 315–321 (2016)
2. Bae, M.M., Bose, B.: Edge disjoint Hamiltonian cycles in k-ary n-cubes and hypercubes. IEEE Trans. Comput. **52**, 1271–1284 (2003)
3. Barth, D., Raspaud, A.: Two edge-disjoint Hamiltonian cycles in the butterfly graph. Inform. Process. Lett. **51**, 175–179 (1994)
4. El-Amawy, A., Latifi, S.: Properties and performance of folded hypercubes. IEEE Trans. Parallel Distrib. Syst. **2**, 31–42 (1991)
5. Hung, R.W.: Embedding two edge-disjoint Hamiltonian cycles into locally twisted cubes. Theoret. Comput. Sci. **412**, 4747–4753 (2011)
6. Hung, R.W.: Constructing two edge-disjoint Hamiltonian cycles and two-equal path cover in augmented cubes. J. Comput. Sci. **39**, 42–49 (2012)
7. Hung, R.W.: The property of edge-disjoint Hamiltonian cycles in transposition networks and hypercube-like networks. Discrete Appl. Math. **181**, 109–122 (2015)
8. Hung, R.W., Chan, S.J., Liao, C.C.: Embedding two edge-disjoint Hamiltonian cycles and two equal node-disjoint cycles into twisted cubes. Lect. Notes Eng. Comput. Sci. **2195**, 362–367 (2012)
9. Hussain, Z.A., Bose, B., Al-Dhelaan, A.: Edge disjoint Hamiltonian cycles in Eisenstein-Jacobi networks. J. Parallel Distrib. Comput. **86**, 62–70 (2015)

10. Lin, T.J., Hsieh, S.Y., Juan, J.S.-T.: Embedding cycles and paths in product networks and their applications to multiprocessor systems. IEEE Trans. Parallel Distrib. Syst. **23**, 1081–1089 (2012)
11. Pai, K.: Embedding three edge-disjoint Hamiltonian cycles into locally twisted cubes. In: Chen, C.-Y., Hon, W.-K., Hung, L.-J., Lee, C.-W. (eds.) COCOON 2021. LNCS, vol. 13025, pp. 367–374. Springer, Cham (2021). https://doi.org/10.1007/978-3-030-89543-3_31
12. Pai, K.J., Wu, R.Y., Peng, S.L., Chang, J.M.: Three edge-disjoint Hamiltonian cycles in crossed cubes with applications to fault-tolerant data broadcasting. J. Supercomput. **79**, 4126–4145 (2023)
13. Petrovic, V., Thomassen, C.: Edge-disjoint Hamiltonian cycles in hypertournaments. J. Graph Theor. **51**, 49–52 (2006)
14. Rowley, R., Bose, B.: Edge-disjoint Hamiltonian cycles in de Bruijn networks. In: Proceedings of the 6th Distributed Memory Computing Conference, pp. 707–709 (1991)
15. Saad, Y., Schultz, M.H.: Topological properties of hypercubes. IEEE Trans. Comput. **37**, 867–872 (1988)
16. Wang, D.: Embedding Hamiltonian cycles into folded hypercubes with faulty links. J. Parallel Distrib. Comput. **61**, 545–564 (2001)
17. Wang, Y., Fan, J., Liu, W., Wang, X.: Embedding two edge-disjoint Hamiltonian cycles into parity cubes. Appl. Mech. Mater. **336–338**, 2248–2251 (2013)

Extraction of Urinary Bubbles from Mongolian Medicine-Related Urine Using Modified VMUNet and Advanced Loss Function

Fucheng Wei[1], Saheya Barintag[2(✉)], Dalai Nashun[3], and Chimedragchaa Chimedtseren[4]

[1] College of Computer Science and Technology, Inner Mongolia Normal University, Hohhot 010022, People's Republic of China

[2] College of Mathematics Science, Inner Mongolia Normal University, Hohhot 010022, People's Republic of China
saheya@imnu.edu.cn

[3] International Mongolian Hospital of Inner Mongolia, Hohhot 010000, People's Republic of China

[4] Institute of Traditional Medicine and Technology of Mongolia, Ulaanbaatar 11000, Mongolia

Abstract. Urine bubble analysis serves as a crucial diagnostic indicator for Heyi syndrome (a key Mongolian medical syndrome) and others. However, current diagnostic methods primarily depend on manual visual examination, which lacks robust automated image analysis methods. To bridge this gap, this paper propose a urine bubble image segmentation approach combining backbone replacement and loss function analysis to facilitate Mongolian medicine diagnosis. Methodologically, we adopt the Vision Mamba U-Net (VMUNet) as the base framework, and explore a backbone replacement strategy to enhance segmentation performance. Specifically, we substitute the original VMUNet backbone with four mainstream segmentation architectures—U^2-Net, U-Net, PSPNet, and DeepLabV3+—and systematically evaluate their performance on the urine bubble segmentation task. For each architecture, we adopt representative loss functions based on their prior relevance to medical image segmentation, including Binary Cross-Entropy (BCE), Dice loss, Tversky loss, Focal Tversky loss, and Boundary loss. This results in a comprehensive set of architecture–loss combinations, enabling a rigorous comparative analysis. Experimental results demonstrate that integrating the DeepLabV3+ architecture with the BCE–Dice loss function yields the best performance in edge restoration, small bubble detection, and background suppression. In particular, it achieves an Intersection over Union (IoU) of 72.88% and an F1 score of 84.31%. These findings validate the potential of this approach for urine bubble extraction and further contribute to expanding the application landscape of artificial intelligence (AI) in traditional and ethnic medical image analysis.

A. Chinchuluun et al. (Eds.): AAIM 2025, LNCS 16379, pp. 86–102, 2026.
https://doi.org/10.1007/978-981-95-5657-1_8

Keywords: Urine bubble segmentation · DeepLabV3+ · VMUNet · U^2-Net · Mongolian medicine diagnosis

1 Introduction

Urinalysis once plays an important role in the diagnosis system of traditional Mongolian medicine, especially in the process of distinguishing cold and heat syndromes. Clinicians will observe various macroscopic urine images as important detection indicators. These indicators usually include urine color, transparency, and floating objects [5]. In addition to these, the presence and size characteristics of bubbles in urine are also considered important physical signs. Compared with modern medicine, which emphasizes the identification of pathological cells, such as the automatic identification and classification of typical particles such as red blood cells, white blood cells, epithelial cells, and crystals, Mongolian medicine places greater emphasis on non-cellular macroscopic characteristics. Medical practitioners typically examine multiple characteristics of urine. Figure 1 presents an illustration of a urine bubble image, while Fig. 2 depicts an example of an image showcasing color and transparency characteristics. However, the current relevant analysis methods rely entirely on manual observation and judgment, which is obviously subjective.

Although manual observation and detection are time-consuming and labor-intensive, they are regarded as a reliable standard for examining the macroscopic features of urine in Mongolian medicine. However, pathologists with varying levels of experience often produce inconsistent reports. Therefore, there is a pressing need for an efficient and accurate automated urine analysis method to guarantee the consistency of test results, thereby facilitating disease diagnosis. In Mongolian medicine theory, diseases are mainly divided into two categories: cold diseases and hot diseases, and bubble size features are given priority in distinguishing Heyi syndrome (cold diseases) from other syndrome. It is crucial to accurately extract the bubble area in urine from urine images, because the subsequent visual feature analysis and classification of Heyi syndrome depend on this basic segmentation result. When dealing with urine bubble images, an ideal segmentation algorithm ought to comprehensively extract all bubble regions relevant to diagnosis while disregarding the container edges and reflective areas. If these non-bubble interference areas are mistakenly passed to the subsequent analysis module, diagnostic bias may occur. In this study, we not only focus on accurately separating bubbles from the background, but also need to reduce the problem of misextraction of areas with similar bubble morphology. In this section, we first explore some challenges in urine bubble extraction and then discuss the main contributions of this paper.

1.1 Challenges in Extracting Bubble Features

Several challenges emerge during the extraction of gas bubbles from urine images:

(a) Densely clustered, irregular bubbles near the edge.

(b) Fewer, well-separated bubbles with larger size.

Fig. 1. Comparison of urine bubble morphology and distribution.

- **Blurred boundaries:** Urine bubbles often show blurred, softened edges, especially when they are attached to the container wall, partially ruptured, or affected by light. This unclear boundary makes it difficult for the segmentation algorithm to accurately locate the bubble area, which can easily result in incomplete boundaries or misidentification.
- **Irregular distribution and morphology:** In contrast to cell-like targets, urine bubbles are distributed very unevenly in images. Some images contain only a few isolated bubbles, while others contain dense overlapping areas. The shapes may also be circular, elliptical, or deformed structures. The extensive morphological diversity poses challenges in generalizing extraction methods relying on rules or geometric priors.
- **Visual similarity to the background:** In most cases, the color or grayscale value of urine bubbles is very similar to that of the urine background, especially under natural light or non-standard shooting conditions, which is prone to low contrast. This visual similarity makes it difficult for the model to distinguish between bubbles and background areas, resulting in missed detection or inaccurate boundary extraction.

(a) Showing different levels of transparency: low transparency on the left and high transparency on the right.

(b) Different colored urine: the urine on the left is white, while the urine on the right is yellow.

Fig. 2. Urine color and transparency feature image samples.

1.2 Contributions

The main contributions of this paper are as follows:

- We applied the deep learning image segmentation model to the Mongolian medicine urine bubble recognition task and proposed an improved image segmentation framework based on the VMUNet framework. This method explored the backbone network structures such as U^2-Net and DeepLabV3+, and strengthened the model's detection and segmentation capabilities for bubbles with blurred boundaries, small sizes, and irregular shapes by replacing the backbone modules. When compared to the original VMUNet architecture, our proposed method has attained remarkable enhancements across multiple evaluation metrics.
- We systematically compared a variety of loss functions and combination strategies, including BCE, Dice, Tversky, Focal Tversky, and Boundary Loss. Experimental results show that the combination of BCE, Dice, Focal Tversky and Boundary has significant advantages in boundary preservation, small object recognition, and category imbalance in the bubble segmentation task.

This paper is organized as follows: Sect. 2 reviews the relevant research progress in urine image segmentation and medical image analysis. Section 3 introduces our overall approach, including the network structure replacement strategy and loss function design. Section 4 describes the image dataset, preprocessing method and enhancement strategy used in the experiment. Section 5 shows the comparative experimental results, visualization performance and ablation analysis of various structure and loss function combinations. Finally, Sect. 6 summarizes the paper.

2 Related Work

Image processing and computer vision techniques have long been widely used in the segmentation of various types of targets in urine microscopy images. Early studies focused on the extraction of red blood cells (RBC), white blood cells (WBC) and epithelial cells. For example, Luo et al. [13] implemented variational segmentation of cell regions based on the Mumford–Shah model [14], and Jiang et al. [8] combined the level set method [15] with the simplified Mumford–Shah function to model cell boundaries. In addition, there are studies that use wavelet transform, edge detection and adaptive thresholding [11] to segment objects such as white blood cells, red blood cells and casts. Cellular neural networks [25], Gabor filters and K-means clustering [24] have also been introduced into this field. However, most of these methods are only applicable to small tiles of 128×128 pixels and have poor adaptability to real full-view urine images. For larger targets (such as casts, epithelial cells) or adherent cell clusters (such as red blood cell clusters, white blood cell clusters), researchers have tried to use variance map histogram method [10], edge enhancement algorithm based on Sobel operator [23], or combined distance transform and watershed algorithm [19] to achieve target extraction. Liancheng et al. [21] proposed a double threshold method based on standard deviation gradient, which can be used to distinguish adherent cells from background areas. These methods have achieved certain results in cell segmentation tasks, but they have obvious shortcomings when dealing with non-cell targets with fuzzy boundaries, complex morphology, and uneven distribution (such as urine bubbles).

Compared with traditional cell targets, urine bubbles have significant differences in visual characteristics. Their boundaries are often blurred or incomplete, and their color and brightness are close to the background urine liquid, making it difficult to distinguish them by fixed thresholds or morphological rules [1]. In addition, the spatial distribution of bubbles is often uneven, sometimes gathering at the edge of the bowl, and sometimes isolated in the central area, which further increases the complexity of segmentation. Existing segmentation methods are mostly designed for targets with clear cell contours and closed structures, and lack adaptability to such small targets with irregular shapes and weak contour continuity [2,4]. Due to these challenges, the use of traditional image processing methods (such as edge detection, grayscale distribution analysis, region growing, etc.) in urine bubble segmentation is prone to edge breakage and background misclassification. Therefore, a model with stronger perception capabilities is needed to achieve accurate segmentation of bubbles and effective suppression of interference areas.

With the rapid development of deep learning technology, semantic segmentation methods based on convolutional neural networks have demonstrated powerful capabilities in the field of medical image processing. From AlexNet [9], VGG [20], GoogLeNet [22], ResNet [6] to DenseNet [7], various network architectures have laid the foundation for the application of deep learning models in classification and segmentation tasks. The fully convolutional neural network (FCN) [12] makes image-level segmentation possible by rewriting the fully connected layer into a convolutional form. Subsequently, U-Net [17] introduced a symmetrical upsampling and skip connection structure based on it, and has been widely used in medical image tasks.

Based on the above work, The paper proposed Vision Mamba U-Net (VMUNet) [18] as a starting point and try to apply it to the segmentation task of urine bubble images. As an encoding-decoding segmentation framework that integrates state space modeling capabilities, VMUNet has strong long-range dependency modeling capabilities and has shown good performance in multiple medical image segmentation tasks. However, in preliminary experiments, we found that VMUNet can hardly effectively segment the bubble area in this task. Most of the model output results are full background or misclassified areas, indicating that its original structure lacks sufficient expression ability for bubble targets with characteristics such as blurred boundaries and irregular shapes. To this end, we try to replace the backbone network of VMUNet with a structure that is better at modeling small targets and edge details. Firstly, U^2-Net [16] is introduced and its multi-scale nested residual module (RSU) is embedded in the original framework to improve the model's adaptability to weak edges and small area targets. This structure has achieved significant improvements in bubble area recognition, but there is still a problem of boundary jitter in some samples. Furthermore, we explored the introduction of DeepLabV3+ [3] as the backbone network and combined it with the Atrous Spatial Pyramid Pooling (ASPP) module to enhance the model's ability to model multi-scale structures. Experimental results demonstrate that this architecture attains the optimal performance across multiple evaluation metrics, particularly in terms of the Intersection over Union (IoU) and F1 scores.

3 Methodology

This paper aims to achieve accurate extraction of tiny bubble areas in urine RGB images. Building upon the VMUNet framework as our baseline architecture, this study conducts a systematic evaluation of how varying backbone replacement strategies and loss function combinations influence segmentation performance. The steps are as follows. We start by substituting the backbone of VMUNet with the multi - scale nested residual structure U^2-Net. Preliminary experimental results indicate that this replacement endows the model with certain advantages in urine bubble recognition, which paves the way for subsequent improvements. To further improve the model's ability to perceive bubble regions of different sizes, we introduce the backbone structure of DeepLabV3+, whose ASPP module can realize multi-scale feature extraction and fusion, effectively addressing the issues of boundary blur and enhancing the recognition of small targets. Finally, we conduct a comprehensive comparison of the impacts of various backbone architectures and combinations of loss functions on the performance of urine bubble segmentation.

3.1 Enhanced Bubble Segmentation via U^2-Net Backbone Integration

In the initial experimental stage, we directly used the original VMUNet structure to segment the bubble targets in urine images. However, the IoU value was only 4.51%. Although the structure has strong long-range modeling capabilities, it performs poorly for small targets, bubble areas with blurred edges and low contrast with the background. The model often outputs all background or mis-segmentation. To this end, we tried to replace the backbone encoder in VMUNet with the multi-scale nested residual module proposed in the U^2-Net structure. The core advantage of U^2-Net is that each of its RSU modules has a complete U-Net substructure nested inside, which can effectively aggregate feature representations of different scales and enhance the model's perception of small targets and fine boundaries. We retain the original decoder structure and overall training process of VMUNet. The encoder part is replaced with a 6-layer RSU module structure as shown in Table 1. Correspondingly, the decoder part consists of Stage5 to Stage1, which uses a symmetrical structure to gradually restore spatial features and fuse jump connection information. The model input is a $256 \times 256 \times 3$ image, and the final output is a binary segmented image.

Table 1. Configuration of RSU Modules in the U^2-Net Encoder (Input: number of input channels, Mid: number of intermediate channels, Output: number of output channels)

Stage	RSU Module	Input	Mid	Output
Stage 1	RSU7	3	32	64
Stage 2	RSU6	64	32	128
Stage 3	RSU5	128	64	256
Stage 4	RSU4	256	128	512
Stage 5	RSU4F	512	256	512
Stage 6	RSU4F	512	256	512

And it is mapped to a probability map through the sigmoid activation function. During the training phase, only the main output d0 is employed for the calculation of the loss function. Meanwhile, the remaining auxiliary outputs are not utilized in the inference process. The structure is shown in Fig. 3.

This strategy significantly improves the model's ability to recognize tiny target bubbles. For example, when the U^2-Net structure and Tversky loss function are used, the IoU index reaches 58.55%. A quantitative comparison analysis with other structures will be conducted in the experimental section.

3.2 Integrating DeepLabV3+ as Backbone to Capture Multi-Scale Context

Although the aforementioned solution of embedding the U^2-Net structure into the VMUNet framework has achieved certain improvements in the task of bubble area segmentation, due to the lack of U^2-Net's ability to model global contextual information, the model still has difficulty in obtaining consistent predictions on complex backgrounds or multi-scale bubble targets. In order to further enhance the model's multi-scale perception and context modeling capabilities, we replaced the VMUNet encoder with the DeepLabV3+ structure in this solution. DeepLabV3+ introduces the Atrous Spatial Pyramid Pooling module based on the original deep neural network, which can effectively capture contextual information under different receptive fields and has stronger modeling capabilities for urine bubble areas with blurred edges or significant scale differences. In this proposed solution, the structure of the VMUNet decoder remains unaltered. Specifically, only the original VM encoder backbone is substituted with the DeepLabV3+ encoder, which is based on the ResNet-50 architecture. To enhance the convergence speed and generalization ability, the pre-trained weights from the ImageNet are employed. The specific replacement strategy is as follows:

1) Feature Extraction: Input images of size $256 \times 256 \times 3$ are passed through a ResNet-50 backbone to generate multi-scale feature maps at different resolutions.

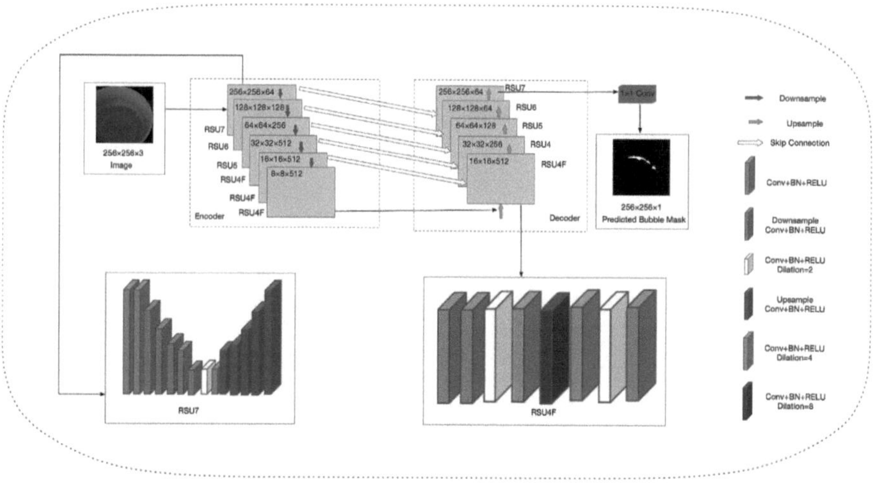

Fig. 3. Structure of the VMUNet segmentation framework enhanced with a U^2-Net backbone. The model processes an input image of size $256 \times 256 \times 3$ through an encoder path generating feature maps of sizes 128×128, 64×64, 32×32, 16×16, and 8×8, followed by a symmetric decoder that upsamples back to $256 \times 256 \times 1$. RSU modules are used in each stage, and skip connections link encoder and decoder layers for detailed boundary preservation.

2) Context Encoding via ASPP: The output feature map from the last ResNet block is fed into the ASPP module, which applies parallel atrous convolutions with dilation rates of 6, 12, and 18, along with global image pooling to enhance receptive field diversity.
3) Decoder and Prediction: The ASPP output is fused with lower-level features and upsampled progressively. A 1×1 convolution layer is used to reduce the channel dimension to 1, followed by a sigmoid activation to produce the final binary segmentation mask.

The whole structure is shown in Fig. 4. This replacement scheme achieved better performance than U^2-Net in subsequent experiments, especially showing higher stability in boundary preservation in bubble areas and small target detection.

3.3 Loss Function Design

The segmentation of bubble regions in urine images faces unique challenges: numerous small instances, blurred boundaries, and irregular contours. These issues demand loss functions that are sensitive to subtle boundary details, robust to class imbalance (since bubbles often occupy a small fraction of the image), and effective at recalling small structures. To address these, we evaluate a set of standard and tailored loss functions, each designed to target specific aspects of the problem. For clarity, we first define key probabilistic metrics used in the loss

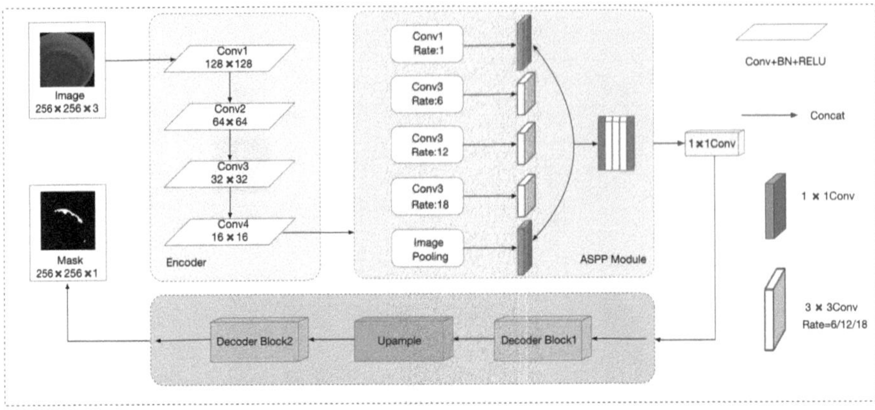

Fig. 4. DeepLabV3+ based segmentation model for urine bubble detection. The input image of size $256 \times 256 \times 3$ is encoded into a 16×16 feature map, processed through the Atrous Spatial Pyramid Pooling (ASPP) module with dilation rates of 6, 12, and 18 plus image-level pooling, and finally upsampled to generate a $256 \times 256 \times 1$ binary mask. The figure depicts all key modules, feature map sizes, and decoding stages for detailed urine bubble extraction.

formulations. For pixel i, let $p_i \in [0, 1]$ denote the predicted foreground probability and $g_i \in \{0, 1\}$ the ground-truth label. We use the following soft counts: $\text{TP} = \sum_i p_i g_i$ (true positives, correctly predicted bubbles), $\text{FP} = \sum_i p_i(1 - g_i)$ (false positives, non-bubble pixels misclassified as bubbles), $\text{FN} = \sum_i (1 - p_i) g_i$ (false negatives, missed bubble pixels), and totals $P = \sum_i p_i$(total predicted foreground), $G = \sum_i g_i$(total ground-truth foreground). Tversky loss is abbreviated as TV and Focal-Tversky as FT. The weights $\alpha, \beta \in [0, 1]$ trade off FP and FN, $\gamma > 0$ is the focusing parameter, and $\epsilon > 0$ is a small constant for numerical stability. All summations are over image pixels unless specified.

BCE: Provides stable gradient flow in early training, critical for initial convergence when learning basic bubble/non-bubble distinctions. It quantifies the discrepancy between predicted probabilities and ground-truth labels:

$$\mathcal{L}_{\text{BCE}} = -\frac{1}{N} \sum_{i=1}^{N} \left[g_i \log p_i + (1 - g_i) \log(1 - p_i) \right]. \tag{1}$$

Dice: Maximizes overlap between predicted and ground-truth bubble regions, directly mitigating class imbalance (common in small bubble instances where background dominates). Its formulation is:

$$\mathcal{L}_{\text{Dice}} = 1 - \frac{2\,\text{TP}}{P + G + \epsilon}. \tag{2}$$

TV: Flexibly adjusts penalties for FP (spurious non-bubble regions) and FN (missed bubbles) via (α, β), making it adaptable to the trade-off between precision and recall in bubble segmentation:

$$\mathcal{L}_{\text{TV}} = 1 - \frac{\text{TP}}{\text{TP} + \alpha\,\text{FP} + \beta\,\text{FN}}. \tag{3}$$

FT: Enhances focus on hard-to-segment regions (e.g., blurred bubble boundaries) by raising the Tversky term to the power γ, amplifying gradients for pixels where predictions are uncertain:

$$\mathcal{L}_{\text{FT}} = \left(1 - \frac{\text{TP}}{\text{TP} + \alpha\,\text{FP} + \beta\,\text{FN}}\right)^{\gamma}. \tag{4}$$

Boundary: Explicitly emphasizes contour alignment by weighting pixel-wise errors with d_i (distance from pixel i to the nearest ground-truth bubble boundary), prioritizing correction of boundary inaccuracies:

$$\mathcal{L}_{\text{Boundary}} = \sum_i d_i \left|p_i - g_i\right|. \tag{5}$$

In the experiment, we combined the above loss functions under the U^2-Net and DeepLabV3+ structures, and constructed multiple training configurations for comparative experiments. The results show that under the U^2-Net and DeepLabV3+ backbone structures, the combination of BCE-Dice is relatively stable in overall performance, with IoU and F1 values reaching 72.88% and 84.31% respectively; and the latter also demonstrates outstanding performance in bubble detection and edge fitting when using other loss functions. This indirectly indicates that the DeepLabV3+ architecture possesses a remarkably robust capability for capturing small targets. Considering comprehensive performance and the ability to depict edge details, the final model employs the DeepLabV3+ backbone architecture and selects BCE-Dice Loss as the optimization objective function. As a result, it achieves the optimal bubble segmentation effect.

4 Dataset and Preprocessing

The dataset employed in this paper was gathered by our laboratory. The images were taken by a CCD industrial camera with a flat LED light source to ensure the illumination consistency and edge clarity of the image acquisition. Each image contains a complete urine sample area, and the container is a standard white porcelain bowl, as shown in Fig. 1. The original resolution of the image is 1028×1024 pixels, and the image scene contains a small number of urine bubbles with different shapes and blurred edges. A total of 59 Mongolian medicine urine diagnosis cases that met strict clinical standards were collected, each case corresponding to 6 to 8 urine images, and the total number of annotated images was 300, and Labelme was leveraged for pixel-level annotation of bubble regions only. Since the bubble edges are often accompanied by fuzzy features, the annotation process was time-consuming, and all masks were manually created to ensure accuracy and consistency. The generated mask is a single-channel binary image, in which the bubble area is the foreground and the rest is the background. Considering the small dataset, we divided the 300 annotated images into training and test sets in a ratio of 8:2. The training images were resized to 256×256 before being input into the model, keeping the RGB three-channel input format, and the corresponding mask image was a single channel. In order to improve the generalization ability of the model under the condition of limited data, a variety of data enhancement strategies were introduced in the training stage, including random angle rotation ($-20°$ to $20°$), random $90°$ rotation and horizontal/vertical flipping. These enhancement strategies enable the model to better adapt to the diversity of bubbles in terms of direction, position, and shape.

It should be emphasized that, considering that bubbles occupy a relatively small proportion in the entire image and the majority of bubbles are distributed in the edge region of the container, there exists an extremely unbalanced pixel distribution between the background and the foreground. This situation not only presents a challenge to the segmentation model but also imposes higher requirements on the design of the loss function and the edge perception capability.

5 Experimental Results Analysis

5.1 Experimental Setup and Training Details

Throughout the training process of all models, unified parameter configurations, as presented below, are employed. The model input image size is uniformly adjusted to 256×256, the number of channels is 3, and the number of output categories is 1. The batch size used for each round of training is 12, the total number of training rounds is set to 1000 rounds, and the number of seeds is set to 42 to ensure the reproducibility of the experiment. All experiments are run on a device numbered GPU, multi-card distributed training is not enabled, and mixed precision training is not enabled. The optimizer uses AdamW, the learning rate is initialized to 1×10^{-3}, and dynamically adjusted through the cosine annealing strategy, the minimum learning rate is set to 1×10^{-5}, and the cycle is 50. During the training process, log information is printed every 20 iterations, the performance is evaluated on the validation set every 30 epochs, and the model weights are saved every 100 rounds of training. In terms of model pre-training strategy, for models based on Segmentation Models PyTorch (SMP) such as DeepLabV3+, PSPNet, and U-Net, the ImageNet pre-trained ResNet encoder is introduced as the backbone; while for native structures such as U^2-Net and VMUNet, they are trained from scratch to ensure the consistency and fairness of the experimental settings.

All experiments were completed on a server equipped with an NVIDIA Tesla P100 12 GB GPU. The software environment includes Python 3.8, PyTorch 1.13, CUDA 12.2, and compatible cuDNN acceleration libraries.

5.2 Comparative Evaluation of Model Architectures and Loss Functions

To comprehensively evaluate the performance of different model architectures and loss function combinations in the task of urine bubble segmentation, we conducted a series of experiments covering mainstream segmentation networks including VMUNet, U^2-Net, UNet, PSPNet, and DeepLabV3+, each trained with commonly used loss functions such as BCE, Dice, Tversky, Boundary, Focal Tversky, and their combinations. All experiments were conducted under consistent datasets and training settings to ensure fairness in comparison.

Table 2. Performance comparison of different model architectures and loss function combinations on the urine bubble segmentation task. Metrics: IoU, F1 Score, Accuracy, Specificity, Sensitivity (all in %).

Model	Loss	IoU	F1 Score	Accuracy	Specificity	Sensitivity
VMUNet	Tversky	**4.51**	8.63	56.45	55.59	95.03
U^2-Net	BCE-Dice	63.71	77.84	98.87	99.53	74.59
	Tversky	**58.55**	73.86	98.58	98.70	92.95
	Focal Tversky	56.20	71.96	98.37	98.41	96.53
UNet	BCE-Dice	5.20	9.88	80.88	81.60	48.43
PSPNet	BCE-Dice	69.77	82.20	99.06	99.55	81.24
DeepLabV3+	BCE-Dice	**72.88**	84.31	99.32	99.67	83.75
	Tversky	**59.94**	74.95	98.27	98.29	97.49
	FT	61.69	76.31	98.42	98.50	95.44
	Boundary	**16.10**	27.72	97.70	100.00	16.10
	FT-Boundary	**68.28**	81.15	98.89	99.15	89.42
	BCE	71.23	83.20	99.13	99.62	80.91

Table 2 summarizes the key performance metrics on the test set for each model-loss configuration, including IoU, F1 score, Accuracy, Specificity, and Sensitivity. The results indicate that U^2-Net and DeepLabV3+ consistently outperform VMUNet and UNet across most loss functions, highlighting the superior modeling capacity of deep residual structures and atrous convolutions for small object segmentation. Notably, DeepLabV3+ paired with BCE-Dice loss achieved the best performance, with an IoU of 72.88% and an F1 score of 84.31%. Although Focal Tversky-Boundary losses enhance the model's sensitivity to boundary regions, they fall slightly behind the BCE-Dice combination in terms of overall metrics. The U^2-Net architecture also performed well with BCE-Dice, producing smooth edge contours and competitive quantitative results. In contrast, the lightweight U-Net architecture showed limited performance, suggesting that its shallow design is insufficient to effectively model the complex morphology of urine bubbles. PSPNet demonstrated strong performance in terms of Accuracy and Sensitivity, making it another viable candidate architecture.

These results provide a solid foundation for the subsequent sections on segmentation visualization and ablation studies, and also clearly guide the selection of the final model configuration.

5.3 Urine Bubble Segmentation Visualization

We selected a representative input image to evaluate the performance of different network architectures and loss function combinations in the task of urine bubble segmentation. As shown in Fig. 5, the visual results of 14 model configurations are presented, clearly illustrating their differences in boundary reconstruction, small object detection, and artifact suppression. Firstly, the combination of VMUNet and Tversky loss (Fig. 5c) generated significant artifacts and failed to successfully identify bubble regions, indicating a clear limitation of the original encoder structure in handling low contrast and small-scale targets. In contrast, the U^2-Net series (Fig. 5d–f) demonstrated superior structural representation and detail recovery. Both the BCE, Dice and Tversky configurations yielded relatively complete and coherent segmentation masks, while the Focal Tversky configuration further enhanced boundary sensitivity, albeit with slight fragmentation observed.

The traditional U-Net model (Fig. 5g) exhibited pronounced distortion, with drifting boundaries and large hollow artifacts, reflecting its limited adaptability to small object segmentation. In comparison, PSPNet (Fig. 5n), equipped with a pyramid pooling module, better preserved multi-scale semantic features, resulting in smooth contour delineation and accurate boundary fitting.

The DeepLabV3+ series (Fig. 5h–m) achieved the most stable and accurate segmentation performance overall. In particular, the BCE-Dice combination (Fig. 5h) closely matched the ground truth in both completeness and boundary precision. The Tversky (Fig. 5i) and FocalTversky (Fig. 5k) variants further improved boundary attention, while the Focal Tversky-Boundary combination (Fig. 5l) enhanced the geometric consistency of contours while maintaining region accuracy. In contrast, using Boundary loss alone (Fig. 5j) led to severe under-segmentation, with nearly no bubble region detected—demonstrating the instability of this loss function when lacking auxiliary region constraints.

Through comprehensive analysis, we conclude that DeepLabV3+ combined with BCE-Dice or Focal Tversky-Boundary loss delivers the best segmentation performance under challenging conditions of blurred boundaries and complex backgrounds. It stands out as the most practically valuable solution proposed in this study.

5.4 Ablation Study

To evaluate the impact of backbone structure and loss function design on the segmentation performance of urinary bubbles, two sets of ablation experiments were conducted: (1) Backbone replacement experiments, which examine performance gains from structural modifications; and (2) Loss function combination experiments, which analyze the effect of different losses on boundary fitting and small-object sensitivity.

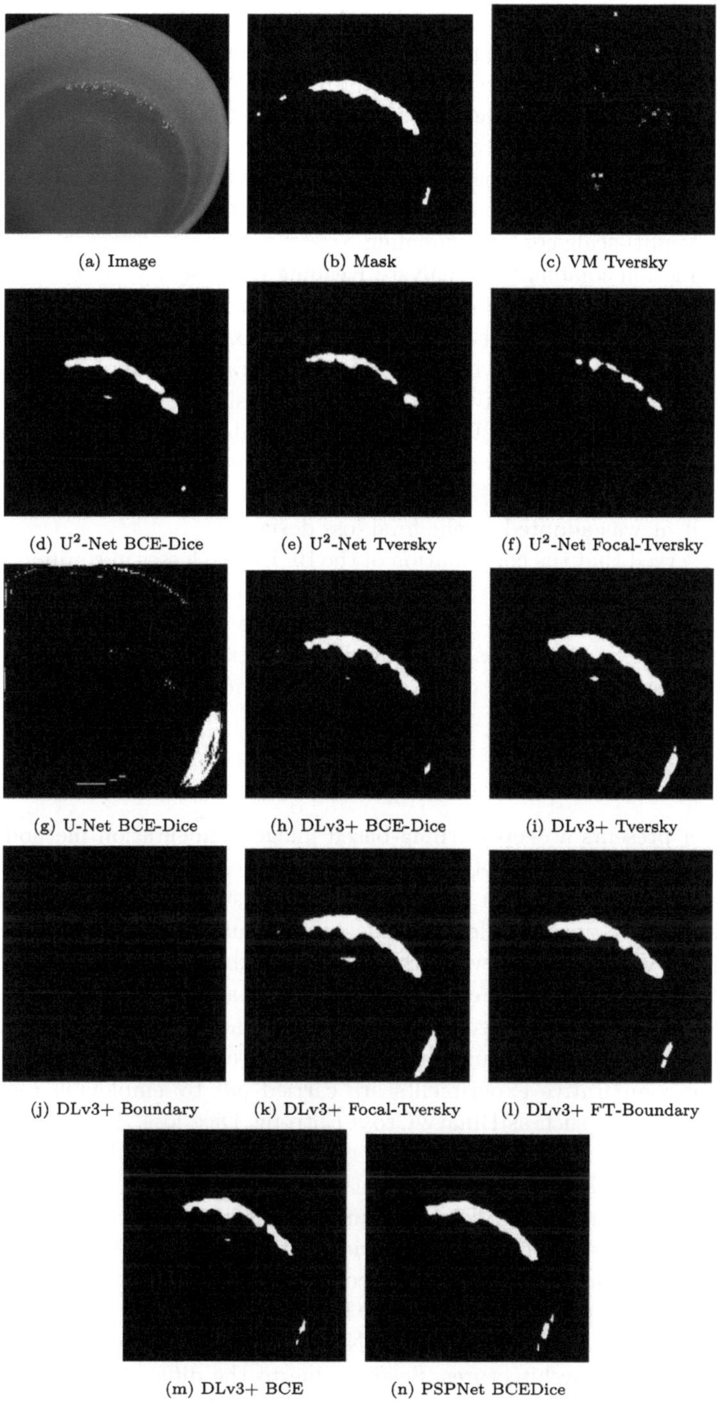

Fig. 5. Qualitative comparison of segmentation results across different model and loss function combinations. Each subfigure represents the predicted mask for the same image input, showing differences in edge alignment, small target detection, and artifact suppression.

Backbone replacement analysis: Based on the VMUNet framework with the original Tversky loss unchanged, we successively replaced its backbone with U^2-Net and DeepLabV3+ to verify the improvement of deep encoding structures. As shown in Table 2, replacing the backbone with U^2-Net improved the IoU from 4.51% to 58.55%, significantly enhancing bubble region detection and boundary restoration. Further replacing it with DeepLabV3+ yielded an IoU of 59.94% and an F1 score of 74.95%, indicating the effectiveness of the ASPP module in enhancing multi-scale context modeling.

Loss function combination analysis: Keeping DeepLabV3+ as the fixed backbone, we evaluated various loss functions including BCE, Dice, Tversky, Focal Tversky, Boundary, and their combinations. The experiment using only Boundary loss showed instability without regional constraints, achieving only 16.10% IoU. Combining Focal Tversky with Boundary loss significantly improved boundary sensitivity, increasing IoU to 68.28%. Ultimately, the combination of the BCE-Dice losses yielded the optimal performance across all evaluated metrics. This combination effectively balanced region completeness and edge smoothness. As a result, it was adopted as the final loss design.

We observed that the combination of the BCE-Dice losses in conjunction with the DeepLabV3+ architecture yielded the optimal overall results. This finding implies that the DeepLabV3+ backbone exhibits a higher degree of responsiveness to effective loss functions. Compared to U^2-Net, DeepLabV3+ proved more suitable for capturing blurry edges and morphologically diverse bubbles.

6 Conclusions

This paper presents a deep learning-based image segmentation method for the automatic extraction of bubble features in Mongolian medicine urine images. To enhance the segmentation performance, an investigation is carried out on the impact of combinations of different model architectures and loss functions. Building upon the VMUNet framework, the original backbone is replaced with U^2-Net, which offers deeper encoding capability, and DeepLabV3+, which incorporates a multi-scale context-aware mechanism. These enhancements aim to improve the detection of small bubbles and targets with blurred boundaries. Furthermore, a series of comparative experiments are carried out by employing representative loss functions, such as Binary Cross Entropy, Dice loss, Tversky loss, Focal Tversky loss, and Boundary loss. The results demonstrate that DeepLabV3+ combined with either BCE-Dice or Focal Tversky-Boundary achieves the best performance in terms of IoU, F1 score, and boundary accuracy, exhibiting strong generalization. Overall, this study provides an effective solution for analysis of Mongolian medicine urine images and contributes to advancing intelligent auxiliary diagnosis in traditional medicine. However, limitations remain: First, the model comparison covers only classic CNN segmentation networks and excludes Transformer-based architectures. Second, due to the difficulty in clinically collecting Mongolian medicine urine images and the need for professional physicians to label them, only 300 labeled images were used for validation, which somewhat limited the evaluation of the model's generalization capabilities. Future research will be further expanded to compare the system performance with representative Transformer-based models, and by collaborating with multiple clinical institutions, the dataset will be expanded to enhance the competitiveness and clinical practical value of the model.

Declarations

Data Availability Statement. This study's privately constructed dataset is not publicly available but can be obtained upon reasonable request with the corresponding author's approval.

Conflicts of Interest. The authors declare no conflicts of interest.

Author contributions. Conceptualization, B. Saheya and Fucheng Wei; methodology, Fucheng Wei, B. Saheya; supervision, Fucheng Wei and Nashundalai; data curation, Nashundalai, Fucheng Wei and Chimedragchaa Chimedtseren; investigation, Nashundalai and Chimedragchaa Chimedtseren; writing—original draft, B. Saheya and Fucheng Wei; writing—review and editing, B. Saheya, Fucheng Wei. All authors have read and agreed to the published version of the manuscript.

References

1. Arbelaez, P., Maire, M., Fowlkes, C., Malik, J.: Contour detection and hierarchical image segmentation. IEEE Trans. Pattern Anal. Mach. Intell. **33**(5), 898–916 (2010)
2. Bui, N.T., Hoang, D.H., Nguyen, Q.T., Tran, M.T., Le, N.: MegaNet: multi-scale edge-guided attention network for weak boundary polyp segmentation. In: Proceedings of the IEEE/CVF Winter Conference on Applications of Computer Vision, pp. 7985–7994 (2024)
3. Chen, L.-C., Zhu, Y., Papandreou, G., Schroff, F., Adam, H.: Encoder-decoder with Atrous separable convolution for semantic image segmentation. In: Ferrari, V., Hebert, M., Sminchisescu, C., Weiss, Y. (eds.) ECCV 2018. LNCS, vol. 11211, pp. 833–851. Springer, Cham (2018). https://doi.org/10.1007/978-3-030-01234-2_49
4. Fan, D.-P., et al.: PraNet: parallel reverse attention network for polyp segmentation. In: Martel, A.L., et al. (eds.) MICCAI 2020. LNCS, vol. 12266, pp. 263–273. Springer, Cham (2020). https://doi.org/10.1007/978-3-030-59725-2_26
5. Fu, S., Bao, G.X., Gao, Y.H.: A brief discussion on the characteristics of urine diagnosis in Mongolian medicine. Chinese J. Ethnic Med. **17**(11), 71–72 (2011). (in Chinese)
6. He, K., Zhang, X., Ren, S., Sun, J.: Deep residual learning for image recognition. In: Proceedings of the IEEE Conference on Computer Vision and Pattern Recognition, pp. 770–778 (2016)
7. Huang, G., Liu, Z., Van Der Maaten, L., Weinberger, K.Q.: Densely connected convolutional networks. In: Proceedings of the IEEE Conference on Computer Vision and Pattern Recognition, pp. 4700–4708 (2017)
8. Jiang, X., Nie, S.: Urine sediment image segmentation based on level set and Mumford-Shah model. In: 2007 1st International Conference on Bioinformatics and Biomedical Engineering, pp. 1028–1030. IEEE (2007)

9. Krizhevsky, A., Sutskever, I., Hinton, G.E.: ImageNet classification with deep convolutional neural networks. Adv. Neural Inf. Process. Syst. **25** (2012)
10. Li, C.Y., Fang, B., Wang, Y., Lu, G.Z., Qian, J.Y., Chen, L.: Automatic detecting and recognition of casts in urine sediment images. In: 2009 International Conference on Wavelet Analysis and Pattern Recognition, pp. 26–31. IEEE (2009)
11. Li, Y.M., Zeng, X.P.: A new strategy for urinary sediment segmentation based on wavelet, morphology and combination method. Comput. Meth. Prog. Biomed. **84**(2-3), 162–173 (2006)
12. Long, J., Shelhamer, E., Darrell, T.: Fully convolutional networks for semantic segmentation. In: Proceedings of the IEEE Conference on Computer Vision and Pattern Recognition, pp. 3431–3440 (2015)
13. Luo, H., Ma, S., Wu, D., Xu, Z.: Mumford-Shah segmentation for microscopic image of the urinary sediment. In: 2007 1st International Conference on Bioinformatics and Biomedical Engineering, pp. 861–863. IEEE (2007)
14. Mumford, D.B., Shah, J.: Optimal approximations by piecewise smooth functions and associated variational problems. Commun. Pure Appl. Math. (1989)
15. Osher, S., Sethian, J.A.: Fronts propagating with curvature-dependent speed: algorithms based on Hamilton-Jacobi formulations. J. Comput. Phys. **79**(1), 12–49 (1988)
16. Qin, X., Zhang, Z., Huang, C., Dehghan, M., Zaiane, O.R., Jagersand, M.: U2-Net: going deeper with nested U-structure for salient object detection. Pattern Recogn. **106**, 107404 (2020)
17. Ronneberger, O., Fischer, P., Brox, T.: U-Net: convolutional networks for biomedical image segmentation. In: Navab, N., Hornegger, J., Wells, W.M., Frangi, A.F. (eds.) MICCAI 2015, Part III 18. LNCS, vol. 9351, pp. 234–241. Springer, Cham (2015). https://doi.org/10.1007/978-3-319-24574-4_28
18. Ruan, J., Li, J., Xiang, S.: VM-UNet: vision Mamba UNet for medical image segmentation. arXiv preprint arXiv:2402.02491 (2024)
19. Shi, Z., Jinlong, Q., Guangjie, Q.: Urinary sediment overlapping cells image segmentation based on combination strategy. In: 2008 International Symposium on Computational Intelligence and Design, vol. 2, pp. 3–7. IEEE (2008)
20. Simonyan, K., Zisserman, A.: Very deep convolutional networks for large-scale image recognition. arXiv preprint arXiv:1409.1556 (2014)
21. Su, L., Jiang, S., Wang, C.: A segmentation method based on standard difference gradient with dual-threshold for urinary sediment visible components. In: 2017 4th International Conference on Information Science and Control Engineering (ICISCE), pp. 94–98. IEEE (2017)
22. Szegedy, C., et al.: Going deeper with convolutions. In: Proceedings of the IEEE Conference on Computer Vision and Pattern Recognition, pp. 1–9 (2015)
23. Wu, Y.H., Kang, M.: A urinary sediment segmentation algorithm of cast and epithelial cells. Appl. Mech. Mater. **457**, 1046–1049 (2014)
24. Zhang, S., Wang, J., Zhao, S., Luan, X.: Urinary sediment images segmentation based on efficient Gabor flters. In: 2007 IEEE/ICME International Conference on Complex Medical Engineering, pp. 812–815. IEEE (2007)
25. Zhang, Z., Xia, S., Duan, H.: Cellular neural network based urinary image segmentation. In: Third International Conference on Natural Computation, ICNC 2007, vol. 2, pp. 285–289. IEEE (2007)

An Inverse Malfatti's Problem

Davaajargal Jargalsaikhan[1,2](✉) and Rentsen Enkhbat[1]

[1] Institute of Mathematics and Digital Technology, Mongolian Academy of Sciences, Ulaanbaatar 13330, Mongolia
renkhbatm@num.edu.mn, davaajargal_j@mas.ac.mn
[2] Department of Applied Mathematics, National University of Mongolia, Ulaanbaatar 14200, Mongolia

Abstract. In this study, we introduce a so-called Inverse Malfatti's problem which is to find a triangle including three non-overlapping circles with the minimum area. This problem will be an inverse version of Malfatti's problem, which dates back to 200 years ago, proposed by the Italian mathematician Malfatti. The Malfatti's problem for the first time has been examined from the point of view of global optimization in [1]. The Inverse Malfatti's problem has also been formulated as the non-convex optimization problem of minimizing a convex function over a non-convex set. Some computational results are provided

Keywords: Malfatti's problem · triangle set · circle · global optimization · Inverse Malfatti's problem

1 Introduction

The problem that has become named as "Malfatti's Problem' dates back to 19th century and proposed by Italian mathematician Gianfrancesco Malfatti (1731-1807) in 1803 [2]. It is concerned with finding three non-overlapping circles with maximum area within a given triangle. There are several research papers on Malfatti's problem [3–8]. In addition, Malfatti's problem is further has been examined from a view point of sphere packing theory [1,9], game theory [10] and multi-objective optimization [11]. The high-dimensional Malfatti's problems were considered in [12,13]. Inspired by the original Malfatti's problem, we introduce Inverse Malfatti's problem which is concerned with finding a triangle with minimum area which includes three non-overlapping circles. This problem is formulated as the convex minimization problem over a non-convex set. To find a solution, we apply a standard penalty method. Numerical experiments were performed by GEKKO in Python Jupyter Notebook and BARON.

2 Methodology

The main methodology of the research is sphere-packing theory. In general, sphere-packing theory deals with packing non-overlapping spheres of the maximum volume in a given set.

A. Chinchuluun et al. (Eds.): AAIM 2025, LNCS 16379, pp. 103–111, 2026.
https://doi.org/10.1007/978-981-95-5657-1_9

Define a sphere with a radius r and center $u^0 \in R^n$.

$$B(u^0, r) = \{x \in R^n | \|u^0 - x\| \leq r\}$$

Denote by $c^1, c^2 \dots c^k$ are the centers of the spheres. Let $r_1, r_1 \dots r_k$ be their corresponding radii, respectively. Optimization problem of maximizing a total volume of k non-overlapping spheres [9] is:

$$max_{(c,r)}\ V = \frac{\pi^{n/2}}{\Gamma(\frac{n}{2}+1)} \sum_{i=1}^{k} r_i^n \tag{1}$$

subject to:

$$\langle a^i, c^j \rangle + r_j \|a^i\| \leq b_i, \;\; i = 1, \dots, m, \;\; j = 1, \dots, k, \tag{2}$$

$$\|c^i - c^j\|^2 \geq (r_i + r_j)^2, \;\; i, j = 1, \dots, k, \;\; i < j. \tag{3}$$

$$r_i \geq 0, \;\; i = 1, \dots, k, \tag{4}$$

here $D = \{x \in \mathbb{R}^n | \langle a^i, x \rangle \leq b_i\}$ is polytope, $a^i \in \mathbb{R}^n, b_i \in \mathbb{R}, \;\; i = \overline{1, m}$. Γ is Euler's gamma function.
If $r_i = r, \;\; i = 1, \dots, k$ then problem (1)-(4) has the following form:

$$max_{(c,r)}\ V = \frac{\pi^{n/2}}{\Gamma(\frac{n}{2}+1)} k r^n \tag{5}$$

subject to:

$$\langle a^i, c^j \rangle + r \|a^i\| \leq b_i, \;\; i = 1, \dots, m \;\; j = 1, \dots, k, \tag{6}$$

$$\|c^i - c^j\|^2 \geq 4r^2, \;\; i, j = 1, \dots, k, \;\; i < j \tag{7}$$

$$r \geq 0. \tag{8}$$

If we now take $n = 2, \;\; m = 3, \;\; k = 3$ in (5)-(8), then it becomes Malfatti's problem [2]

2.1 Malfatti's Problem and Convex Maximization

Malfatti's problem is to pack three non-overlapping disks in a triangle set with the maximum area. Assume that a triangle set $D \subset R^2$ is given by

$$D = \{x \in R^2 | \langle a^i, x \rangle \leq b_i\}, \;\; a^i \in R^2, b_i \in R, i = 1, 2, 3, \tag{9}$$

here, $\langle , \rangle$ denotes the scalar product of two vectors in R^2, and $\|\cdot\|$ is the Euclidean norm, $a^i \nparallel a^j, \; i \neq j; \; i, j = 1, 2, 3$.

Then condition $B(u^0, r) \subset D$ is characterized by the following conditions:

$$\langle a^i, u \rangle + x_3 \|a^i\| \leq b_i, i = 1, 2, 3, \tag{10}$$

$$\langle a^i, v \rangle + x_6 \|a^i\| \leq b_i, i = 1, 2, 3, \tag{11}$$

$$\langle a^i, p \rangle + x_9 \|a^i\| \leq b_i, i = 1, 2, 3. \tag{12}$$

Denote by $u(x_1, x_2)$, $v(x_4, x_5)$ and $p(x_7, x_8)$ centers of three disks inscribed in a triangle set D given by (9). Let x_3,x_6 and x_9 be their corresponding radii. Malfatti's problem is to find three non-overlapping circles with maximum area in a given triangle [1]. This problem was formulated in [1] as the following convex maximization problem and solved numerically by a global optimization method.

$$\max f = \pi(x_3^2 + x_6^2 + x_9^2), \tag{13}$$
$$\langle a^i, u\rangle + x_3\|a^i\| \leq b_i, i = 1,2,3, \tag{14}$$
$$\langle a^i, v\rangle + x_6\|a^i\| \leq b_i, i = 1,2,3, \tag{15}$$
$$\langle a^i, p\rangle + x_9\|a^i\| \leq b_i, i = 1,2,3, \tag{16}$$
$$(x_4 - x_1)^2 + (x_5 - x_2)^2 \geq (x_3 + x_6)^2, \tag{17}$$
$$(x_7 - x_1)^2 + (x_8 - x_2)^2 \geq (x_3 + x_9)^2, \tag{18}$$
$$(x_7 - x_4)^2 + (x_8 - x_5)^2 \geq (x_6 + x_9)^2, \tag{19}$$
$$x_3 \geq 0, x_6 \geq 0, x_9 \geq 0. \tag{20}$$

2.2 Inverse Malfatti's Problem and Non-Convex Minimization

Denote $B_1(u^*, x_3^*)$, $B_2(v^*, x_6^*)$ and $B_3(p^*, x_9^*)$ as Malfatti's circles found by solving the problem (13)-(20). Let $A_1(y_1, y_2)$, $A_2(y_3, y_4)$ and $A_3(y_5, y_6)$ be vertices of the triangle of minimum area. Here $u^* = (u_1^*, u_2^*)$, $v^* = (v_1^*, v_2^*)$ and $p^* = (p_1^*, p_2^*)$

Definition 1. *The problem of finding a triangle with minimum area including Malfatti's circles is called the Inverse Malfatti's problem.*

The conditions of the circles $B_1(u^*, x_3^*)$, $B_2(v^*, x_6^*)$ and $B_3(p^*, x_9^*)$ which belong to a triangle are the following:

$$\langle \bar{a}^i, u^*\rangle + x_3^*\|\bar{a}^i\| \leq \bar{b}_i, i = 1,2,3, \tag{21}$$
$$\langle \bar{a}^i, v^*\rangle + x_6^*\|\bar{a}^i\| \leq \bar{b}_i, i = 1,2,3, \tag{22}$$
$$\langle \bar{a}^i, p^*\rangle + x_9^*\|\bar{a}^i\| \leq \bar{b}_i, i = 1,2,3. \tag{23}$$

where $\bar{a}^1 = (y_4 - y_2, y_1 - y_3)$, $\bar{a}^2 = (y_6 - y_4, y_3 - y_5)$, $\bar{a}^3 = (y_2 - y_6, y_5 - y_1)$, $\bar{b}_1 = (y_4 - y_2)y_1 + (y_1 - y_3)y_2$, $\bar{b}_2 = (y_6 - y_4)y_3 + (y_3 - y_5)y_4$, and $\bar{b}_3 = (y_2 - y_6)y_5 + (y_5 - y_1)y_6$. Then the area of the triangle can be easily computed as follows:

$$S = \frac{1}{2} \det \begin{pmatrix} y_1 & y_2 & 1 \\ y_3 & y_4 & 1 \\ y_5 & y_6 & 1 \end{pmatrix} \tag{24}$$

Here, *det* is determinant of matrix. Then Inverse Malfatti's problem is:

$$\min S, \tag{25}$$
$$\langle \bar{a}^i, u^*\rangle + x_3^*\|\bar{a}^i\| \leq \bar{b}_i, i = 1,2,3, \tag{26}$$
$$\langle \bar{a}^i, v^*\rangle + x_6^*\|\bar{a}^i\| \leq \bar{b}_i, i = 1,2,3, \tag{27}$$
$$\langle \bar{a}^i, p^*\rangle + x_9^*\|\bar{a}^i\| \leq \bar{b}_i, i = 1,2,3. \tag{28}$$

Problem (25)-(28) can be expressed in the following form:

$$\min\{\frac{1}{2}\det\begin{pmatrix} y_1 & y_2 & 1 \\ y_3 & y_4 & 1 \\ y_5 & y_6 & 1 \end{pmatrix}\}$$

$$u_1^*(y_4-y_2)+u_2^*(y_1-y_3)+x_3^*\sqrt{(y_4-y_2)^2+(y_1-y_3)^2} \le (y_4-y_2)y_1+(y_1-y_3)y_2$$
$$v_1^*(y_4-y_2)+v_2^*(y_1-y_3)+x_6^*\sqrt{(y_4-y_2)^2+(y_1-y_3)^2} \le (y_4-y_2)y_1+(y_1-y_3)y_2$$
$$p_1^*(y_4-y_2)+p_2^*(y_1-y_3)+x_9^*\sqrt{(y_4-y_2)^2+(y_1-y_3)^2} \le (y_4-y_2)y_1+(y_1-y_3)y_2$$
$$u_1^*(y_6-y_4)+u_2^*(y_3-y_5)+x_3^*\sqrt{(y_6-y_4)^2+(y_3-y_5)^2} \le (y_6-y_4)y_3+(y_3-y_5)y_4$$
$$v_1^*(y_6-y_4)+v_2^*(y_3-y_5)+x_6^*\sqrt{(y_6-y_4)^2+(y_3-y_5)^2} \le (y_6-y_4)y_3+(y_3-y_5)y_4$$
$$p_1^*(y_6-y_4)+p_2^*(y_3-y_5)+x_9^*\sqrt{(y_6-y_4)^2+(y_3-y_5)^2} \le (y_6-y_4)y_3+(y_3-y_5)y_4$$
$$u_1^*(y_2-y_6)+u_2^*(y_5-y_1)+x_3^*\sqrt{(y_2-y_6)^2+(y_5-y_1)^2} \le (y_2-y_6)y_5+(y_5-y_1)y_6$$
$$v_1^*(y_2-y_6)+v_2^*(y_5-y_1)+x_6^*\sqrt{(y_2-y_6)^2+(y_5-y_1)^2} \le (y_2-y_6)y_5+(y_5-y_1)y_6$$
$$p_1^*(y_2-y_6)+p_2^*(y_5-y_1)+x_9^*\sqrt{(y_2-y_6)^2+(y_5-y_1)^2} \le (y_2-y_6)y_5+(y_5-y_1)y_6 \quad (29)$$

Thus, Inverse Malfatti's problem is a non-convex minimization problem over a nonconvex set.

2.3 Penalty Method

For solving problem (25)-(28), we use the penalty method. In our case, we consider inequality constrained optimization problems of the following form:

$$\min_{y\in Y\subset R^n} f(y) \quad (30)$$

subject to :

$$g_i(y) \le 0; \quad i=1,\ldots,m, \quad (31)$$

where $f(y)$ is:

$$f(y)=\tfrac{1}{2}\det\begin{pmatrix} y_1 & y_2 & 1 \\ y_3 & y_4 & 1 \\ y_5 & y_6 & 1 \end{pmatrix}$$

and $g(y)$ are:

$$g_1(y) = u_1^*(y_4 - y_2) + u_2^*(y_1 - y_3) + x_3^*\sqrt{(y_4 - y_2)^2 + (y_1 - y_3)^2} - (y_4 - y_2)y_1 + (y_1 - y_3)y_2 \leq 0$$
$$g_2(y) = v_1^*(y_4 - y_2) + v_2^*(y_1 - y_3) + x_6^*\sqrt{(y_4 - y_2)^2 + (y_1 - y_3)^2} - (y_4 - y_2)y_1 + (y_1 - y_3)y_2 \leq 0$$
$$g_3(y) = p_1^*(y_4 - y_2) + p_2^*(y_1 - y_3) + x_9^*\sqrt{(y_4 - y_2)^2 + (y_1 - y_3)^2} - (y_4 - y_2)y_1 + (y_1 - y_3)y_2 \leq 0$$
$$g_4(y) = u_1^*(y_6 - y_4) + u_2^*(y_3 - y_5) + x_3^*\sqrt{(y_6 - y_4)^2 + (y_3 - y_5)^2} - (y_6 - y_4)y_3 + (y_3 - y_5)y_4 \leq 0$$
$$g_5(y) = v_1^*(y_6 - y_4) + v_2^*(y_3 - y_5) + x_6^*\sqrt{(y_6 - y_4)^2 + (y_3 - y_5)^2} - (y_6 - y_4)y_3 + (y_3 - y_5)y_4 \leq 0$$
$$g_6(y) = p_1^*(y_6 - y_4) + p_2^*(y_3 - y_5) + x_9^*\sqrt{(y_6 - y_4)^2 + (y_3 - y_5)^2} - (y_6 - y_4)y_3 + (y_3 - y_5)y_4 \leq 0$$
$$g_7(y) = u_1^*(y_2 - y_6) + u_2^*(y_5 - y_1) + x_3^*\sqrt{(y_2 - y_6)^2 + (y_5 - y_1)^2} - (y_2 - y_6)y_5 + (y_5 - y_1)y_6 \leq 0$$
$$g_8(y) = v_1^*(y_2 - y_6) + v_2^*(y_5 - y_1) + x_6^*\sqrt{(y_2 - y_6)^2 + (y_5 - y_1)^2} - (y_2 - y_6)y_5 + (y_5 - y_1)y_6 \leq 0$$
$$g_9(y) = p_1^*(y_2 - y_6) + p_2^*(y_5 - y_1) + x_9^*\sqrt{(y_2 - y_6)^2 + (y_5 - y_1)^2} - (y_2 - y_6)y_5 + (y_5 - y_1)y_6 \leq 0$$

The penalty method is expressed as follows:

$$p(y) = \sum_{i=1}^{m} (\max(g_i(y), 0))^2, \tag{32}$$

here $p(y)$ is continuous, Y is a compact set in R^n and $\{C_k\}_{k=1}^{\infty}$-is a positive sequence and it grows monotonously. The convergence of the penalty function approach is given by the following theorem.

Theorem 1. [14]. *Assume that* $y^k = \arg\min_{y \in Y}\{f(y) + C_k p(y)\}$ *and* $0 < C_1 < C_2 < \ldots < C_k$. *Let* $\hat{y}$ *be an arbitrary limit point of* $\{y^k\}_{k=1}^{\infty}$. *Then,* $\hat{y}$ *is a local solution of the problem (30)-(31).*

The algorithm of the penalty method consists of the following steps:

Algorithm PM: Penalty method

Step 1: Parameters $c_1, m, k \in N, \varepsilon_y > 0$ are given and $k = 1, c_1 = 10, m = 2, \varepsilon_y = 10^{-6}$

Step 2: Solve the minimization problem

$$\min_{y \in Y} L(y, C_k) = \min_{y \in Y} (f(y) + C_k p(y)) \tag{33}$$

get a local solution $y^k = \arg\min L(y, C_k), y \in Y$.

Step 3: $u = y^k, \quad k = k + 1, \quad C_k = mC_{k-1}$

Step 4: Solve the following minimization problem

$$\min_{y \in Y} L(y, C_k) = \min_{y \in Y} (f(y) + C_k p(y)) \tag{34}$$

to obtain another a local solution $y^k = \arg\min L(y, C_k), y \in Y$.

Step 5: If $\|u - y^k\| \leq \varepsilon_y$ then algorithm stops and solution y^k is a local minimum (or a stationary point) of problem (30)-(31). Otherwise goto Step 3.

3 Numerical Results

For solving the Inverse Malfatti's problem, Algorithm PM has been coded in Python Jupyter notebook and implemented numerically. For numerical implementation, we take a triangle with vertices $A(0,0)$, $B(3,4)$, $C(8,6)$, and an area of ($S_{ABC}=7$) considered in [1].We consider three cases of Malfatti circles found by a local search method in [1], which are Case 1 - Case 3 (Table 1 and Figs. 1, 2 and 3)

Table 1. Numerical results for Malfatti's problem [1]

Case	x_1	x_2	x_3	x_4	x_5	x_6	x_7	x_8	x_9	f
Case 1	4.49	4.02	0.52	2.58	2.58	0.51	3.43	3.43	0.68	3.2
Case 2	4.49	4.02	0.52	4.27	3.9	0.15	3.43	3.43	0.68	2.42
Case 3	3.43	3.43	0.68	3.24	2.62	0.14	2.58	2.58	0.51	2.39

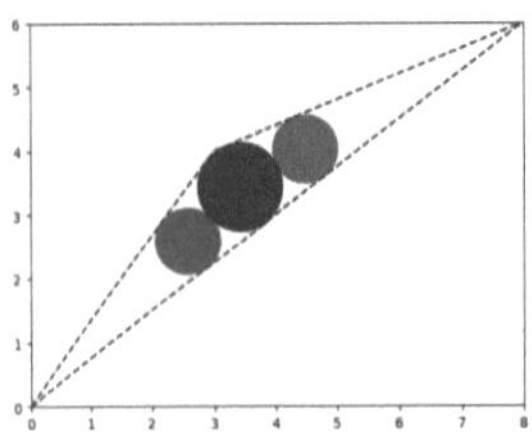

Fig. 1. Case 1

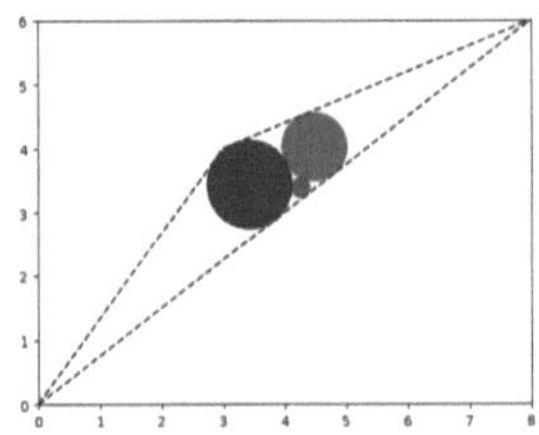

Fig. 2. Case 2

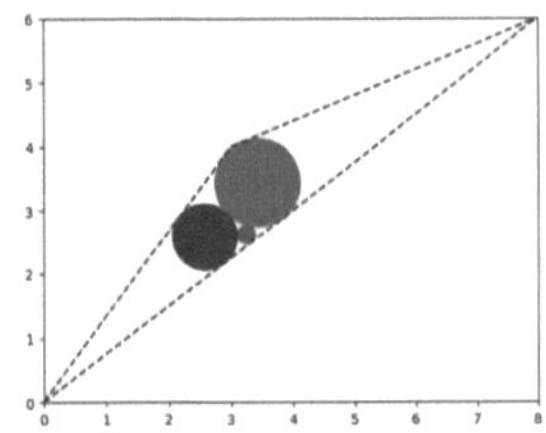

Fig. 3. Case 3

For case 1, we can write the following Inverse Malfatti's problem:

$$\min\{\frac{1}{2}\det\begin{pmatrix} y_1 & y_2 & 1\\ y_3 & y_4 & 1\\ y_5 & y_6 & 1\end{pmatrix}\}$$

$$\begin{aligned}
&4.49(y_4-y_2)+4.02(y_1-y_3)+0.52\sqrt{(y_4-y_2)^2+(y_1-y_3)^2}\le(y_4-y_2)y_1+(y_1-y_3)y_2\\
&2.58(y_4-y_2)+2.58(y_1-y_3)+0.51\sqrt{(y_4-y_2)^2+(y_1-y_3)^2}\le(y_4-y_2)y_1+(y_1-y_3)y_2\\
&3.43(y_4-y_2)+3.43(y_1-y_3)+0.68\sqrt{(y_4-y_2)^2+(y_1-y_3)^2}\le(y_4-y_2)y_1+(y_1-y_3)y_2\\
&4.49(y_6-y_4)+4.02(y_3-y_5)+0.52\sqrt{(y_6-y_4)^2+(y_3-y_5)^2}\le(y_6-y_4)y_3+(y_3-y_5)y_4\\
&2.58(y_6-y_4)+2.58(y_3-y_5)+0.51\sqrt{(y_6-y_4)^2+(y_3-y_5)^2}\le(y_6-y_4)y_3+(y_3-y_5)y_4\\
&3.43(y_6-y_4)+3.43(y_3-y_5)+0.68\sqrt{(y_6-y_4)^2+(y_3-y_5)^2}\le(y_6-y_4)y_3+(y_3-y_5)y_4\\
&4.49(y_2-y_6)+4.02(y_5-y_1)+0.52\sqrt{(y_2-y_6)^2+(y_5-y_1)^2}\le(y_2-y_6)y_5+(y_5-y_1)y_6\\
&2.58(y_2-y_6)+2.58(y_5-y_1)+0.51\sqrt{(y_2-y_6)^2+(y_5-y_1)^2}\le(y_2-y_6)y_5+(y_5-y_1)y_6\\
&3.43(y_2-y_6)+3.43(y_5-y_1)+0.68\sqrt{(y_2-y_6)^2+(y_5-y_1)^2}\le(y_2-y_6)y_5+(y_5-y_1)y_6
\end{aligned}\quad(35)$$

We have obtained the following results: $y^*=(6.15,4.62,2.77,4.74,1.42,1.06)$ and the total minimum area S = 5.66.

For case 2, we have the following Inverse Malfatti's problem:

$$\min\{\tfrac{1}{2}\det\begin{pmatrix} y_1 & y_2 & 1 \\ y_3 & y_4 & 1 \\ y_5 & y_6 & 1 \end{pmatrix}\}$$

$$\begin{aligned}
4.49(y_4-y_2)+4.02(y_1-y_3)+0.52\sqrt{(y_4-y_2)^2+(y_1-y_3)^2} &\le (y_4-y_2)y_1+(y_1-y_3)y_2\\
4.27(y_4-y_2)+3.9(y_1-y_3)+0.15\sqrt{(y_4-y_2)^2+(y_1-y_3)^2} &\le (y_4-y_2)y_1+(y_1-y_3)y_2\\
3.43(y_4-y_2)+3.43(y_1-y_3)+0.68\sqrt{(y_4-y_2)^2+(y_1-y_3)^2} &\le (y_4-y_2)y_1+(y_1-y_3)y_2\\
4.49(y_6-y_4)+4.02(y_3-y_5)+0.52\sqrt{(y_6-y_4)^2+(y_3-y_5)^2} &\le (y_6-y_4)y_3+(y_3-y_5)y_4\\
4.27(y_6-y_4)+3.9(y_3-y_5)+0.15\sqrt{(y_6-y_4)^2+(y_3-y_5)^2} &\le (y_6-y_4)y_3+(y_3-y_5)y_4\\
3.43(y_6-y_4)+3.43(y_3-y_5)+0.68\sqrt{(y_6-y_4)^2+(y_3-y_5)^2} &\le (y_6-y_4)y_3+(y_3-y_5)y_4\\
4.49(y_2-y_6)+4.02(y_5-y_1)+0.52\sqrt{(y_2-y_6)^2+(y_5-y_1)^2} &\le (y_2-y_6)y_5+(y_5-y_1)y_6\\
4.27(y_2-y_6)+3.9(y_5-y_1)+0.15\sqrt{(y_2-y_6)^2+(y_5-y_1)^2} &\le (y_2-y_6)y_5+(y_5-y_1)y_6\\
3.43(y_2-y_6)+3.43(y_5-y_1)+0.68\sqrt{(y_2-y_6)^2+(y_5-y_1)^2} &\le (y_2-y_6)y_5+(y_5-y_1)y_6 \qquad (36)
\end{aligned}$$

Computational algorithm provides the following solutions:
$y^* = (6.06, 5.22, 2.17, 3.67, 3.76, 2.19)$ and total minimum area $S = 4.1$.
For case 3, we have the problem:

$$\min\{\tfrac{1}{2}\det\begin{pmatrix} y_1 & y_2 & 1 \\ y_3 & y_4 & 1 \\ y_5 & y_6 & 1 \end{pmatrix}\}$$

$$\begin{aligned}
3.43(y_4-y_2)+3.43(y_1-y_3)+0.0.68\sqrt{(y_4-y_2)^2+(y_1-y_3)^2} &\le (y_4-y_2)y_1+(y_1-y_3)y_2\\
3.24(y_4-y_2)+2.62(y_1-y_3)+0.14\sqrt{(y_4-y_2)^2+(y_1-y_3)^2} &\le (y_4-y_2)y_1+(y_1-y_3)y_2\\
2.58(y_4-y_2)+2.58(y_1-y_3)+0.51\sqrt{(y_4-y_2)^2+(y_1-y_3)^2} &\le (y_4-y_2)y_1+(y_1-y_3)y_2\\
3.43(y_6-y_4)+3.43(y_3-y_5)+0.68\sqrt{(y_6-y_4)^2+(y_3-y_5)^2} &\le (y_6-y_4)y_3+(y_3-y_5)y_4\\
3.24(y_6-y_4)+2.62(y_3-y_5)+0.14\sqrt{(y_6-y_4)^2+(y_3-y_5)^2} &\le (y_6-y_4)y_3+(y_3-y_5)y_4\\
2.58(y_6-y_4)+2.58(y_3-y_5)+0.51\sqrt{(y_6-y_4)^2+(y_3-y_5)^2} &\le (y_6-y_4)y_3+(y_3-y_5)y_4\\
3.43(y_2-y_6)+3.43(y_5-y_1)+0.68\sqrt{(y_2-y_6)^2+(y_5-y_1)^2} &\le (y_2-y_6)y_5+(y_5-y_1)y_6\\
3.24(y_2-y_6)+2.62(y_5-y_1)+0.14\sqrt{(y_2-y_6)^2+(y_5-y_1)^2} &\le (y_2-y_6)y_5+(y_5-y_1)y_6\\
2.58(y_2-y_6)+2.58(y_5-y_1)+0.51\sqrt{(y_2-y_6)^2+(y_5-y_1)^2} &\le (y_2-y_6)y_5+(y_5-y_1)y_6 \qquad (37)
\end{aligned}$$

with the solution of $y^* = (4.68, 3.51, 2.85, 4.54, 1.35, 1.01)$ and total minimum area $S = 4$. The numerical results are summarized in the Table 2 and Figs. 4, 5 and 6.

Table 2. Numerical results of Inverse Malfatti's problem for Case 1 - Case 3

Case	y_1	y_2	y_3	y_4	y_5	y_6	S	BARON(t)	GEKKO(t)	Algorithm PM(t)
Case $\hat{1}$	6.15	4.62	2.77	4.74	1.42	1.06	5.66	2.3 s	3.9 s	5.1 s
Case $\hat{2}$	6.06	5.22	2.17	3.67	3.76	2.19	4.1	2.2 s	3.8 s	4.7 s
Case $\hat{3}$	4.68	3.51	2.85	4.54	1.35	1.01	4	2.4 s	3.9 s	4.9 s

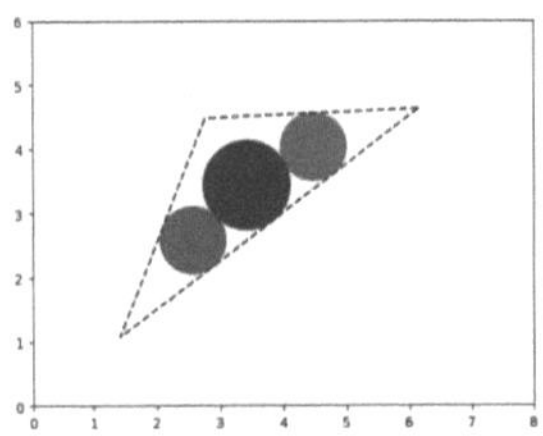

Fig. 4. Case $\hat{1}$

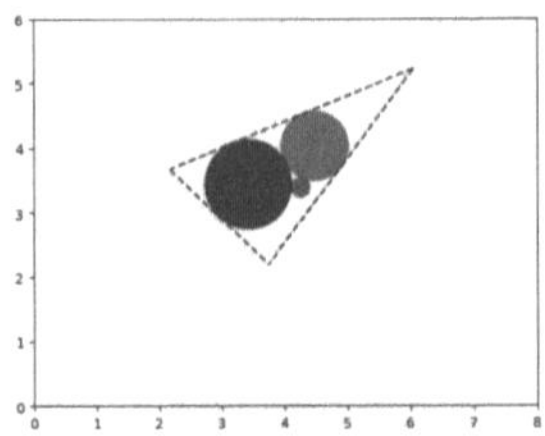

Fig. 5. Case $\hat{2}$

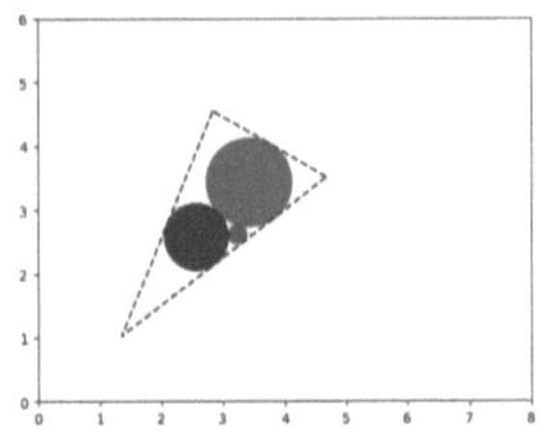

Fig. 6. Case $\hat{3}$

4 Conclusion

In this paper, for the first time, we introduce so-called Inverse Malfatti's problem based on the Malfatti's problem. Geometrically, the Inverse Malfatti's problem is to find a triangle with the minimum area which contains three non-overlapping circles. The problem belongs to a class of global optimization. Using penalty method, we have made an attempt to solve the problem numerically. The numerical results were conducted on GEKKO in a Python Jupyter notebook and BARON. Similarly, we can formulate the inverse sphere packing problem of finding a polytope with the minimum volume which contains non-overlapping spheres. It will be examined in a next paper.

References

1. Enkhbat, R.: Global optimization approach to Malfatti's problem. J. Glob. Optim. **65**, 33–39 (2016). https://doi.org/10.1007/s10898-015-0372-6
2. Malfatti, C.: Memoria sopra una problema stereotomico. Memoria di Matematica e di Fisica della Societa ttaliana della Scienze **10**(1), 235–244 (1803)
3. Zalgaller, V.A., Los, G.A.: The solution of Malfatti's problem. J. Math. Sci. **72**(6), 3163–3177 (1994)
4. Andreatta, M., Bezdek, A., Boronski, J.P.: The problem of malfatti: two centuries of debate. Math. Intell. **33**(1), 72–76 (2011)
5. Lob, H., Richmond, H.W.: On the solutions of the Malfatti problem for a triangle. Proc. London Math. **2**(30), 287–301 (1930)
6. Goldberg, M.: On the original Malfatti's problem. Math. Mag. **40**(5), 241–247 (1967)
7. Los, G. A.: Malfatti's optimization problem [in Russian]. Dep. Ukr.NIINTI July 5 (1967)
8. Gabai, H., Liban, E.: On Goldbergs inequality associated with the Malfatti's problem. Math. Mag. **41**(5), 251–252 (1967)
9. Enkhbat, R.: Convex maximization formulation of general sphere packing problem. Bull. Irkutsk State Univ. Ser. Math. **31**, 142–149 (2020)
10. Enkhbat, R.,Battur.G.: Generalized Nash equilibrium problem based on Malfatti's problem. Numerical Algebra, Control Optim. **11**(2), 209–220 (2021)
11. Enkhbat, R., Battur, G.: Multi-objective optimization approach to malfatti's problem. Contrib. Game Theory Manage. **14**, 82–90 (2021)

12. Enkhbat, R.,Barkova.M.V., Strekalovsky.A.S.: Solving Malfatti's high dimensional problem by global optimization. Numerical Algebra, Control Optim. **6**(2), 153–160 (2021)
13. Enkhbat, R.,Barkova.M.V.: Global search method for solving malfatti's four-circle problem. J. Irkutsk State Univ. (Ser. Math.) **15**, 38–49 (2016)
14. Nocedal, J., Wright, S.J.: Numerical Optimization. Springer (1999). https://doi.org/10.1007/978-0-387-40065-5

On a Bilevel Optimization Model of Interaction Between One Importer and Several Exporters on the Coal Market

Andrei V. Orlov[1(✉)] and Batbileg Sukhee[2]

[1] Matrosov Institute for System Dynamics and Control Theory of SB of RAS, Irkutsk, Russia
anor@icc.ru

[2] National University of Mongolia, Ulaanbaatar, Mongolia
batbileg@seas.num.edu.mn

Abstract. This paper addresses the interactions in the economic system with one importer and several exporters of the homogeneous product on the example of the international coal market. The importer sets prices and additional tariffs for each supplier country for a number of coal resources types. The suppliers determine the quantity of the supplied product, in consideration of their costs and supply capabilities. To optimize such a system, we propose a hierarchical (bilevel) model where the upper-level (the importer) aims to minimize the costs of each product type, taking into account the satisfaction of its needs. And the goal of each exporter is to maximize its profit, given the established prices and tariffs, as well as its own production costs and supply capacity. To develop numerical methods for the considered problem, we begin by reformulating the bilevel problem as a single-level nonconvex optimization problem, substituting the lower level with its optimality conditions. Next, we apply the Exact Penalization Theory along with the Global Search Theory (GST) to tackle the transformed problem. GST involves expressing nonconvex functions as DC representations (that is, as the difference of two convex functions) and constructing both local and global search methods tailored to the distinct features of the examined problem.

Keywords: Coal market · Import-export of coal resources · Prices and tariffs · Bilevel optimization model · KKT-approach · Exact Penalization Theory · Global Search Theory · Local search · Global Search Scheme

1 Introduction

Coal is a very important energy asset. If we consider the world balance of energy consumption by sources of its production, coal still occupies a leading position in energy generation. The global coal trade remains a critical component of the energy and industrial sectors, with China serving as the dominant importer

A. Chinchuluun et al. (Eds.): AAIM 2025, LNCS 16379, pp. 112–124, 2026.
https://doi.org/10.1007/978-981-95-5657-1_10

because of its vast energy demands and metallurgical industry (it accounts for almost half of the world's consumption), even though it is among the top five countries in the world in terms of own coal reserves. This demand is attracting an ever-increasing number of potential suppliers, while existing ones are trying to protect their shares in this growing market by investing in capacity extension. Thus, it is vital to determine the strategies of the major coal exporters to China and evaluate possible relations within this system.

Now, the main countries supplying coal to the Chinese market are Australia, Russia and Mongolia. Among the coal resources types, thermal coal (energy coal), coking coal and anthracite are usually distinguished. In 2023–2024, the international coal market saw important changes. Total global coal imports by China stabilized at approximately 310–320 million tons annually, reflecting both policy shifts towards cleaner energy and steady domestic demand [1,2]. Key coal types showed the following distinct trade patterns.

1. *Thermal Coal.* China imported about 175–185 million tons annually during this period, primarily sourced from a limited number of countries. The supplier distribution in the Chinese market was roughly: Australia—38–40%, Russia—28–30%, Mongolia—15–17%, and other countries—around 13–14% [1–3]. Due to evolving geopolitical tensions and supply chain adjustments, Mongolia and Russia increased their shares, often supplying thermal coal blends with relatively lower emissions.
2. *Coking Coal.* Critical for steel production, coking coal imports reached approximately 85–90 million tons per year. Australia remained the largest supplier with around 48–50% market share, followed by Russia at 30–32%, Mongolia near 10%, and others covering about 8–10% [1–3].
3. *Anthracite.* Although a smaller segment (roughly 12–15 million tons yearly), anthracite imports to China were mainly from Russia (around 70%), with Mongolia and other countries accounting for the remainder [3].

The remaining 25–35 million tons correspond primarily to lignite (brown coal) and other less significant coal types. Overall, the total coal import market shares in China for 2023–2024 can be summarized as: Australia—about 38%, Russia—28%, Mongolia—14%, and other countries—20% [3].

China's strategic diversification aims to reduce dependence on any single exporter amid international trade uncertainties. Russia leverages geographic proximity and expanded rail infrastructure, Mongolia capitalizes on resource development and export corridors, while Australia benefits from established maritime logistics and premium coal quality [1,2]. As for China's current tariff policies on the coal market, it aims to balance domestic coal industry interests and import diversification. The current tariff framework encourages import substitution away from Australian coal towards Russian and Mongolian sources, aligning with China's broader geopolitical and economic strategies [1,2].

It is therefore worthwhile to attempt building a mathematical model of these interactions that enables quantitative analysis of supply-demand balance, price stabilization, and the impact of geopolitical or tariff policy changes. Taking into account the unequal relationship between the importer and exporters, it seems reasonable to use a bilevel optimization framework to this end.

Bilevel optimization is a widely used tool for modeling different economic systems with unequal statuses of the participants (refer to Chap. 20 in [4]). In such systems, certain elements have their own goals that generally do not coincide with the goal of the system as a whole. Also, an analysis of such systems does not conform to the regular optimization theory, because the interaction between the elements of the system makes the concept of optimality more complex. The research of different bilevel problems from the perspective of developing efficient numerical methods remains a significant topic in the contemporary theory and methods of Mathematical Optimization [4].

The bilevel problems in the classical statement [4,5] represent optimization problems, which, side by side with ordinary equality and inequality constraints, include a constraint described as an optimization subproblem (lower level problem). At the lower level, one or several players, depending on the upper level, can be modeled. In the latter case, these players should be independent of each other (and make decisions simultaneously and separately). Then we can assume that only one aggregated player acts on the lower level.

So, the proposed bilevel model addresses an economic system composed of the one importing country of a number types of a homogenous product (coal) and several exporting countries. At the same time, the importer has its own potential to provide itself with coal, but it is insufficient for fully covering its own needs and/or is quite expensive in terms of the cost price. The issue of providing the exporters themselves with coal resources is not considered in the model (we assume that only the surplus is offered for sale). Each exporter at the lower level wants to maximize its profit by determining the volume of supplies depending on the prices (and additional tariffs) specified at the upper level. Additional consideration of tariffs in the economic model has recently become quite popular (see, e.g., [6]) and is dictated by contemporary geopolitical reality. Assume that we know the parameters of the quadratic cost functions for supplier countries. In turn, the importer aims to minimize the costs of the produced and purchased product of each type, taking into account the satisfaction of its needs.

The proposed bilevel model features quadratic objective functions and linear constraints at both levels, and is analyzed using a well-known two-stages approach. First, we use the fact that a bilevel optimization problem (in the global sense) can be equivalently formulated as a nonconvex optimization problem [4,5] by utilizing the Karush-Kuhn-Tucker conditions for the convex lower level alongside a penalty method. Second, we apply the Global Search Theory (GST), introduced by A.S. Strekalovsky, to address the resulting nonconvex formulation [7–9]. Significantly, the GST open the way for developing effective numerical algorithms for different classes of bilevel problems [10–15]. This approach has shown its workability not only for an array of benchmark problems of differing scale and complexity [10–12,14], but also for practical bilevel applications in telecommunications and energy sectors [13,15]. In contrast to classical global optimization approaches, GST acknowledges advancements in contemporary convex optimization, making extensive use of efficient numerical methods and software for solving subproblems with convex structure during global search algorithms construction [7,9]. Based on the cited results concerning an application of the GST for studying bilevel optimization problems, we can expect a positive outcome of our research of the proposed coal bilevel model both from theoretical and computational aspects.

The paper is organized in the following way. Section 2 describes the problem formulation and its bilevel model. Section 3 provides the transformation of the bilevel problem into a single-level problem with a DC objective function. In Sect. 4, the special procedure of the Local Search is developed for this problem. Section 5 addresses the Global Search Scheme and Sect. 6 concludes the paper.

2 The Model and its Bilevel Formulation

Let I be the set of exporting countries (suppliers), $I = \{1, 2, \ldots, M\}$; J be the set of coal resource types under consideration, $J = \{1, 2, \ldots, N\}$; Ω be the set of time periods (e.g., months), $\Omega = \{1, 2, \ldots, T\}$; D_{jt} ($j \in J$, $t \in \Omega$) stands for the known needs of the importer for each type of product in each period. In addition, in each period, the importer sets the prices at which it is ready to purchase each type of product from each supplier (p_{ijt}, $i \in I$, $j \in J$, $t \in \Omega$) and the so-called discriminatory tariffs [6] for the supplied coal resources (z_{ijt}, $i \in I$, $j \in J$, $t \in \Omega$). Next, x_{ijt} is the quantity of each product type supplied by exporter i in each time period ($i \in I$, $j \in J$, $t \in \Omega$). The goal of each exporter is to maximize its profit, given the established prices and tariffs, as well as its own production costs and supply capacity. If the costs are quadratic (c_{ij}, q_{ij} are the corresponding coefficients), then the problem of exporter i has the following form ($x^i := (x_{ijt})_{j \in J, t \in \Omega}$, $i \in I$):

$$\left.\begin{array}{l} \sum\limits_{j \in J} \sum\limits_{t \in \Omega} (p_{ijt} x_{ijt} - z_{ijt} x_{ijt} - c_{ij} x_{ijt} - q_{ij} x_{ijt}^2) \uparrow \max\limits_{x^i}, \\ 0 \le x_{ijt} \le \bar{x}_{ij}, \quad j \in J,\ t \in \Omega, \qquad \sum\limits_{t \in \Omega} x_{ijt} \le R_{ij}, \quad j \in J. \end{array}\right\} \qquad (\mathcal{FP}_i(p, z))$$

Here the profit of the ith exporter is defined as the revenue from the sale of supplied coal minus the tariffs (with the linear dependence on the quantity of coal supplied) minus the quadratic production cost for each type of the product. $\bar{x}_{ij}$ are the bounds from above on the supplies of each exporter for each product type (unchanged for all t), R_{ij} are the total constraints on the supplies for all time periods under consideration, $p := (p_{ijt})_{i \in I, j \in J, t \in \Omega}$, $z := (z_{ijt})_{i \in I, j \in J, t \in \Omega}$. Since the problems $(\mathcal{FP}_i(p, z))$, $i \in I$, are independent of each other, the aggregated lower level problem can be formulated as the combination of all exporters' problems ($x := (x_{ijt})_{i \in I, j \in J, t \in \Omega}$):

$$\left.\begin{array}{c} \sum\limits_{i \in I} \sum\limits_{j \in J} \sum\limits_{t \in \Omega} (p_{ijt} x_{ijt} - z_{ijt} x_{ijt} - c_{ij} x_{ijt} - q_{ij} x_{ijt}^2) \uparrow \max\limits_{x}, \\ 0 \le x_{ijt} \le \bar{x}_{ij}, \quad i \in I,\ j \in J,\ t \in \Omega, \\ \sum\limits_{t \in \Omega} x_{ijt} \le R_{ij}, \quad i \in I,\ j \in J. \end{array}\right\} \qquad (\mathcal{FP}(p, z))$$

Further, let y_{jt} $(j \in J,\ t \in \Omega)$ be the importer's variables that determine its own production ($\bar{y}_j$ are the production capabilities for each product type), then the upper-level problem of minimizing the costs of each product type, in consideration of the satisfaction of needs (D_{jt}), is written as follows (c_j^y, q_j^y are the coefficients of the importer's cost function):

$$\left.\begin{array}{c} \sum\limits_{j\in J}\sum\limits_{t\in\Omega}(c_j^y y_{jt} + q_j^y y_{jt}^2) + \sum\limits_{i\in I}\sum\limits_{j\in J}\sum\limits_{t\in\Omega}(p_{ijt}x_{ijt} - z_{ijt}x_{ijt}) \downarrow \min\limits_{p,z,y,x} \\ p_{ijt} \geq 0,\ \ z_{ijt} \geq 0,\ \ i \in I,\ j \in J,\ t \in \Omega, \\ 0 \leq y_{jt} \leq \bar{y}_j,\ \ j \in J,\ t \in \Omega, \\ y_{jt} + \sum\limits_{i\in I} x_{ijt} = D_{jt},\ \ j \in J,\ t \in \Omega; \qquad x \in \mathrm{Sol}(\mathcal{FP}(p, z)). \end{array}\right\} \qquad (\mathcal{BP})$$

The key constraint in the model is the equality constraint that ensures the total quantity of coal produced and supplied is equal to the needs of the importer ($\forall j \in J$ and $\forall t \in \Omega$). It is important to highlight that we are dealing with an optimistic bilevel formulation, meaning the decisions at the upper level are assumed to be agreed with the actions of the lower level [5,10]. We can use such a formulation because the exporters' problems are strictly convex with respect to their variables if the variables of the upper level are fixed. Hence, the whole lower level problem ($\mathcal{FP}(p, z)$) has a unique solution for any fixed (p, z). Consequently, we can disregard the guaranteed approach to bilevel problems and the objective function $F(\cdot)$ can be minimized with respect to both upper- and lower-level variables simultaneously [5,10].

Analyzing the model's structure, the upper-level objective function is quadratic with respect to variable y and contains several bilinear terms, while the lower-level objective function represents a general, although separable, quadratic form also featuring bilinear terms. All constraints at both levels are linear. The solution methodology for the bilevel problem ($\mathcal{BP}$) in this paper builds on the findings from [13–15]. Papers [13,15] addressed specific quadratic applied bilevel problems, and new solution methods for general quadratic bilevel problems were developed and tested in [14].

Following the approach described in [13–15], the solution process consists of: reducing the original bilevel problem to a single-level nonconvex optimization, where nonconvexity appears only in the objective function; constructing an explicit decomposition of this nonconvex function as the difference of two convex functions (DC decomposition); and exploiting the A. S. Strekalovsky's Global Search Theory (GST), which involves two principal steps:

1) A specialized Local Search Method (LSM) tailored to the specific structure of the problem in question [7,9].
2) A procedure based on Alexander Strekalovsky's Global Optimality Conditions (GOCs) [7–9], allowing further improvement of the solution obtained through the LSM [7,10,13–15].

3 Transformation of the Bilevel Problem and DC Approach to the Reduced One

To transform the bilevel problem $(\mathcal{BP})$ to a single-level one, a standard technique is often used, which consists in replacing the lower level problem (which is convex with respect to its variables) with its optimality conditions (it is commonly referred to as the KKT-approach) [5,10–15].

In order to do it, let us introduce the following notations: $n{:=}N{\cdot}T$, $x^i{:=}(x_{ijt})_{j\in J,t\in\Omega}\in I\!R^n$, $p^i{:=}(p_{ijt})_{j\in J,t\in\Omega}\in I\!R^n$, $z^i{:=}(z_{ijt})_{j\in J,t\in\Omega}\in I\!R^n$, $c^i := (\underbrace{c_{i1},\dots,c_{i1}}_{T},\underbrace{c_{i2},\dots,c_{i2}}_{T},\dots,\underbrace{c_{iN},\dots,c_{iN}}_{T})^\top \in I\!R^n$, $Q^i := 2\cdot diag(\underbrace{q_{i1},\dots,q_{i1}}_{T},\underbrace{q_{i2},\dots,q_{i2}}_{T},\dots,\underbrace{q_{iN},\dots,q_{iN}}_{T}) \in I\!R^{n\times n}$.

Next, let E_n denote the identity matrix of dimension n, $B_1=(E_n,-E_n)^\top\in I\!R^{2n\times n}$, $b_1^i=(\bar{x}^i,0_n)^\top\in I\!R^{2n}$, $i\in I$ (here $\bar{x}^i := (\underbrace{\bar{x}_{i1},\dots,\bar{x}_{i1}}_{T},\dots,\underbrace{\bar{x}_{iN},\dots,\bar{x}_{iN}}_{T})^\top$); $b_2=(R_{ij})\in I\!R^{MN}$; $e_T = (1,1,...,1)\in I\!R^T$.

With this, we can express the problem $(\mathcal{FP}(p,z))$ in matrix-vector formulation:

$$\sum_{i\in I}\Big(\langle p^i-z^i-c^i,x^i\rangle-\frac{1}{2}\langle x^i,Q^ix^i\rangle\Big)\uparrow\max_x,\qquad Bx\le b,\qquad (\mathcal{FP}(p,z))$$

$$\text{where } B=\begin{pmatrix} B_1 & 0 & \dots & 0\\ 0 & B_1 & \dots & 0\\ \dots & \dots & \dots & \dots\\ 0 & 0 & \dots & B_1\\ e_T & 0 & \dots & 0\\ 0 & e_T & \dots & 0\\ \dots & \dots & \dots & \dots\\ 0 & 0 & \dots & e_T \end{pmatrix}\in I\!R^{m\times nM},\quad b=(b_1^1,...,b_1^M,b_2)^\top\in I\!R^m,$$

$m := 2nM+MN$ stands for the total number of constraints.

The KKT-conditions for $(\mathcal{FP}(p,z))$ can then be described as the following system, where $v\in I\!R^m$ is the vector of Lagrange multipliers corresponding to the imposed constraints:

$$\langle b-Bx,v\rangle=0,\qquad v\ge 0_m,\qquad c-p+z+v^\top B+Qx=0_{nM},$$

where $c=(c^1,...,c^M)^\top\in I\!R^{nM}$, $p=(p^1,...,p^M)^\top\in I\!R^{nM}$, $z=(z^1,..., z^M)^\top\in I\!R^{nM}$, $Q=diag(Q^1,...,Q^M)$.

This allows us to construct an equivalent single-level problem:

$$\langle p-z,x\rangle+\langle c^y,y\rangle+\frac{1}{2}\langle y,Q^yy\rangle\downarrow\min_{p,z,y,x,v},\qquad (1)$$

$$
\begin{array}{lr}
p \geq 0_{nM}, \quad z \geq 0_{nM}, \quad Bx \leq b, \quad v \geq 0_m, & (2)\\
0 \leq y_{jt} \leq \bar{y}_j, \qquad\qquad j \in J,\ t \in \Omega, & (3)\\
y_{jt} + \sum_{i\in I} x_{ijt} = D_{jt}, \qquad\qquad j \in J,\ t \in \Omega, & (4)\\
c - p + z + v^\top B + Qx = 0_{nM}, & (5)\\
\langle b - Bx, v\rangle = 0, & (6)
\end{array}
$$

where $y := (y_{jt})_{j\in J, t\in\Omega} \in I\!R^n$,
$c^y := (\underbrace{c_1^y, \ldots, c_1^y}_{T}, \underbrace{c_2^y, \ldots, c_2^y}_{T}, \ldots, \underbrace{c_N^y, \ldots, c_N^y}_{T})^\top \in I\!R^n$,
$Q^y := 2 \cdot diag(\underbrace{q_1^y, \ldots, q_1^y}_{T}, \underbrace{q_2^y, \ldots, q_2^y}_{T}, \ldots, \underbrace{q_N^y, \ldots, q_N^y}_{T}) \in I\!R^{n\times n}$.

By aggregating the linear constraints (2)–(5) into the convex set S, we arrive at the following nonconvex optimization problem with a quadratic-bilinear objective function (1) and the bilinear equality constraint:

$$
\langle p - z, x\rangle + \langle c^y, y\rangle + \frac{1}{2}\langle y, Q^y y\rangle \downarrow \min_{p,z,y,x,v}, \ (p, z, y, x, v) \in S, \ \langle b - Bx, v\rangle = 0. \qquad (\mathcal{P})
$$

Drawing on the results from [4,5,13,14], it can be guaranteed that any global optimum to the problem $(\mathcal{P})$ is also a global solution to the original bilevel problem $(\mathcal{BP})$. Moreover, it is a known fact that a bilinear function, being nonconvex, can be represented as the difference between two convex functions (i.e., it pertains to the class of DC functions) [10–14,16]. Global Search Theory [10] provides effective techniques for handling optimization problems involving such nonconvex structures.

The main difficulty lies in the double nonconvexity present in $(\mathcal{P})$, as both its objective function and some of its constraints are nonconvex. This feature can be called "the second degree of nonconvexity" [7,13,15].

To reformulate the problem $(\mathcal{P})$ so that nonconvexity arises only in the objective function, the exact penalty approach is applied [8,9,13,14,17], introducing a family of σ-parametric problems (note that for any feasible $(x, v) \in Pr_{xv}(S) \triangleq \{(x, v) \mid \exists(p, z, y) : (p, z, y, x, v) \in S\}$ the inequality $\langle b - Bx, v\rangle \geq 0$ holds):

$$
\left.\begin{array}{l}
\Phi(p, z, y, x, v) = \langle p - z, x\rangle + \langle c^y, y\rangle + \frac{1}{2}\langle y, Q^y y\rangle + \\
\quad +\sigma\langle b - Bx, v\rangle \downarrow \min, \quad (p, z, y, x, v) \in S,
\end{array}\right\} \qquad (\mathcal{P}(\sigma))
$$

where σ is a penalty parameter ($\sigma > 0$). For a fixed σ, this problem becomes a DC minimization problem [7] with a convex feasible set.

Additionally, if there exists $\hat{\sigma} > 0$ such that equality (6) is satisfied by a solution $(p(\hat{\sigma}), z(\hat{\sigma}), y(\hat{\sigma}), x(\hat{\sigma}), v(\hat{\sigma}))$ of $(\mathcal{P}(\hat{\sigma}))$, then for all $\sigma > \hat{\sigma}$, the function $\langle b - Bx(\sigma), v(\sigma)\rangle$ remains zero, and the corresponding 5-tuple continues to be a solution for $(\mathcal{P})$.

Classical penalization theory implies that if for some value of σ the 5-tuple $(p(\sigma), z(\sigma), y(\sigma), x(\sigma), v(\sigma)) \in \text{Sol}(\mathcal{P}(\sigma))$, and $(p(\sigma), z(\sigma), y(\sigma), x(\sigma), v(\sigma))$ is feasible in the problem $(\mathcal{P})$, i.e. $(p(\sigma), z(\sigma), y(\sigma), x(\sigma), v(\sigma)) \in S$ and $\langle b - Bx(\sigma), v(\sigma)\rangle = 0$, then $(p(\sigma), z(\sigma), y(\sigma), x(\sigma), v(\sigma))$ is a global solution to the problem $(\mathcal{P})$ [8,17]. Additionally, if there exists $\hat{\sigma} > 0$ such that equality (6) is satisfied by a solution $(p(\hat{\sigma}), z(\hat{\sigma}), y(\hat{\sigma}), x(\hat{\sigma}), v(\hat{\sigma}))$ to the problem $(\mathcal{P}(\hat{\sigma}))$, then for all $\sigma > \hat{\sigma}$, the function $\langle b - Bx(\sigma), v(\sigma)\rangle$ remains zero, and the 5-tuple $(p(\sigma), z(\sigma), y(\sigma), x(\sigma), v(\sigma))$ continues to be a solution to the problem $(\mathcal{P})$.

An important aspect in the application of exact penalization theory is the existence of a threshold value $\hat{\sigma} > 0$ for the penalty parameter, such that $\langle b - Bx(\sigma), v(\sigma)\rangle = 0$ for all $\sigma \geq \hat{\sigma}$.

Since the objective function of the problem $(\mathcal{P})$ possesses the Lipschitz property [17] with respect to all variables, we can guarantee the existence of such a threshold $\hat{\sigma} > 0$, as well as the equivalence of the problems $(\mathcal{P})$ and $(\mathcal{P}(\sigma))$ in the sense of the following equality (see [8,14]): $\text{Sol}(\mathcal{P}) = \text{Sol}(\mathcal{P}(\sigma)) \;\; \forall \sigma > \hat{\sigma}$.

Therefore, these established relationships between the problems $(\mathcal{P}(\sigma))$ and $(\mathcal{P})$ allow us to search for a global solution to just a single problem $(\mathcal{P}(\sigma))$, where $\sigma > \hat{\sigma}$, to find an optimistic solution to the bilevel problem $(\mathcal{P})$, instead of solving a whole sequence of problems $(\mathcal{P}(\sigma))$ with $\sigma \to +\infty$ as typically required in classical penalization theory [17].

Now, we can proceed to apply the Global Search Theory to the problem $(\mathcal{P}(\sigma))$ with a fixed $\sigma > \hat{\sigma}$. As indicated earlier, the first step is to express the objective function $\Phi(\cdot)$ explicitly as the difference of two convex functions. This can be accomplished using the standard DC decomposition for the scalar product: $\langle a_1, a_2\rangle = \frac{1}{4}\|a_1 + a_2\|^2 - \frac{1}{4}\|a_1 - a_2\|^2$ (see, e.g., [10]). Hence,

$$\Phi(p, z, y, x, v) = g(p, z, y, x, v) - h(p, z, x, v), \tag{7}$$

where $g(p, z, y, x, v) = \frac{1}{4}\|x + p\|^2 + \frac{1}{4}\|x - z\|^2 + \frac{1}{4}\sigma\|v - Bx\|^2 + \langle c^y, y\rangle + \frac{1}{2}\langle y, Q^y y\rangle + \sigma\langle b, v\rangle$, $h(p, z, x, v) = \frac{1}{4}\|x - p\|^2 + \frac{1}{4}\|x + z\|^2 + \frac{1}{4}\sigma\|v + Bx\|^2$.

It can be easily demonstrated that the functions $g(\cdot)$ and $h(\cdot)$ are convex with respect to all their variables (refer to [10,13,14]). The difference in the number of variables in these functions leads us to the so-called property of non-full (non-full-sized) nonconvexity (see [16]). Furthermore, according to the previously mentioned GST, it is necessary to "lighten" the function $h(\cdot)$, which determines the basic nonconvexity in the problem under consideration. Therefore, convex with respect to variable y terms of the function $\Phi(\cdot)$ have been included into the function $g(\cdot)$. Next, it is necessary to develop a local search procedure that takes into account the specifics of the current problem.

4 Local Search Method

It should be emphasized that in the problem $(\mathcal{P}(\sigma))$, nonconvexity is present in the objective function only. However, in contrast to the problems considered in

[13,14], here the source of nonconvexity is not just a single bilinear term, but three types of bilinear terms, each involving different groups of variables: (p, x), (z, x), and (x, v) (see also [15]). Despite this complexity, we can still apply the standard local search idea for problems with bilinear structures [10,12–16]. This involves dividing the variables into two groups and alternately solving subproblems with the variables of one group fixed. In our situation, unlike the conditions in [13,15], the feasible region of $(\mathcal{P}(\sigma))$ cannot be split into two completely independent sets. This leads to a more elaborate construction of the local search procedure, as described in [14]. Nevertheless, each subproblem still represents a convex optimization problem. A key point is that this approach does not rely on the DC decomposition of the objective function $\Phi(\cdot)$.

Practically, if we fix groups (y, x) and (p, z, v) (denoted by tildes), then the problem $(\mathcal{P}(\sigma))$ can be reduced to two convex problems with respect to (p, z, v) and (y, x), respectively:

$$\left.\begin{array}{c} \langle \widetilde{x}, p\rangle - \langle \widetilde{x}, z\rangle + \sigma\langle b - B\widetilde{x}, v\rangle \downarrow \min\limits_{p,z,v}, \\ p \geq 0_{nM}, \quad z \geq 0_{nM}, \quad v \geq 0_m, \\ v^\top B - p + z = -Q\widetilde{x} - c. \end{array}\right\} \qquad (\mathcal{LP}_{pzv}(\widetilde{y}, \widetilde{x}))$$

$$\left.\begin{array}{c} \frac{1}{2}\langle y, Q^y y\rangle + \langle c^y, y\rangle + \langle \widetilde{p} - \widetilde{z}, x\rangle - \sigma\langle \widetilde{v}^\top B, x\rangle \downarrow \min\limits_{y,x}, \\ 0 \leq y_{jt} \leq \bar{y}_j, \quad j \in J,\ t \in \Omega; \qquad Bx \leq b, \\ y_{jt} + \sum\limits_{i \in I} x_{ijt} = D_{jt}, \quad j \in J,\ t \in \Omega; \\ Qx = \widetilde{p} - \widetilde{z} - c - \widetilde{v}^\top B. \end{array}\right\} \qquad (\mathcal{QP}_{yx}(\widetilde{p}, \widetilde{z}, \widetilde{v}))$$

The first problem is an LP-problem, while the second problem is a convex quadratic problem.

The local search scheme can be organized, for example, as follows. Assume we have an initial point $(p_0, z_0, y_0, x_0, v_0) \in S$, and $\{\rho_s\}$ denotes a sequence of tolerances for convex problem solving at each iteration.

YX-procedure

Step 0. Set $s := 0,\ \ y^s = y_0,\ \ x^s := x_0$.

Step 1. Find an approximate ρ_s-solution $(p^{s+1}, z^{s+1}, v^{s+1})$ to the LP-problem $(\mathcal{LP}_{pzv}(y^s, x^s))$ by some linear programming method.

Step 2. Using some well-known tool, find a ρ_s-solution (y^{s+1}, x^{s+1}) to the quadratic problem $(\mathcal{QP}_{yx}(p^{s+1}, z^{s+1}, v^{s+1}))$.

Step 3. Set $s := s + 1$ and move to **Step 1**. #

According to the results in [10,13,14], this local search procedure converges to a so-called *critical point* in the problem $(\mathcal{P}(\sigma))$. Such a point is a partly global solution with respect to the two groups of variables (p, z, v) and (y, x). This definition of a critical point is especially convenient when performing a global search for problems with a bilinear structure, and it is closely related to classical stationarity definitions [10,12–14,16].

For the computational experiment, we also use the concept of an *approximate critical point* [10,13,14], introducing an additional **Step 2a** that includes a stopping criterion check:

$$\Phi(p^{s+1}, z^{s+1}, y^s, x^s, v^{s+1}) - \Phi(p^{s+1}, z^{s+1}, y^{s+1}, x^{s+1}, v^{s+1}) \leq \tau, \tag{8}$$

where $\tau \geq 0$ is a tolerance. It can be proven (see [10,12–14] for details) that if the stopping criterion (8) is satisfied at iteration s $(s > 1)$ of the YX-procedure, then the 5-tuple $(p^{s+1}, z^{s+1}, y^s, x^s, v^{s+1})$ is a partly global $(\tau + 2\rho_s)$-solution with respect to (p, z, v) and a partly global $(\tau + \rho_s)$-solution with respect to (y, x). Thus, $(p^{s+1}, z^{s+1}, y^s, x^s, v^{s+1})$ is an approximate critical point.

To obtain a feasible point (from the set S) to start the local search procedure described above (and the Global Search Scheme from the next section), it is necessary to solve the convex projection problem for any unfeasible point $(p^0, z^0, y^0, x^0, v^0)$ in the problem $(\mathcal{P}(\sigma))$ onto the convex set S (see [10,13–15]).

5 Global Search Scheme

Obviously, in nonconvex problems, the described local search method generally does not allow one to find a global solution (even in small-dimensional nonconvex cases), since it can only produce critical (stationary) points. Therefore, according to the Global Search Theory [7], the second stage for finding global solutions to nonconvex problems is to develop a procedure for escaping from critical points based on the Global Optimality Conditions (GOCs) [7–9]. Such an approach is possible because of the DC structure of the nonconvex problem $(\mathcal{P}(\sigma))$ when σ is fixed (see DC decomposition (7)).

These conditions use the idea of linearization focused on the basic nonconvexity of the problem at hand and are related to classical optimality conditions. In addition, GOCs possess the following algorithmic (constructive) property: if they are not satisfied, then it is possible to construct a feasible point that improves the value of the objective function compared even to the current critical point [7–9].

In essence, GOCs reduce the study of the nonconvex problem $(\mathcal{P}(\sigma))$ to analyzing a family of convex linearized problems, which depend on $(\widetilde{p}, \widetilde{z}, \widetilde{x}, \widetilde{v})$:

$$g(p, z, y, x, v) - \langle \nabla h(\widetilde{p}, \widetilde{z}, \widetilde{x}, \widetilde{v}), (p, z, x, v) \rangle \downarrow \min_{p,z,y,x,v} \quad (p, z, y, x, v) \in S, \tag{$\mathcal{PL}(\widetilde{p}, \widetilde{z}, \widetilde{x}, \widetilde{v})$}$$

where $\nabla_p h = 1/2(p-x)$; $\nabla_z h = 1/2(z+x)$; $\nabla_x h = 1/2(2x-p+z+\sigma B^\top(v+Bx))$; $\nabla_v h = 1/2(\sigma(v + Bx))$. Each such problem has a quadratic structure and can be solved using standard convex optimization methods and appropriate computational software.

Based on the GOC's constructive property mentioned above, one can propose a special procedure for escaping critical points obtained at the local search stage (see, for instance, [14]). The Global Search Scheme (GSS), which incorporates this procedure, can be formulated written for the current problem $(\mathcal{P}(\sigma))$ (with fixed σ) in the following way (see also [13–15]).

Given a starting point $(p^0, z^0, y^0, x^0, v^0) \in S$, numerical sequences $\{\tau_k\}$, $\{\delta_k\}$ (where $\tau_k,\ \delta_k > 0,\ k = 1, 2, ...$ with $\tau_k \downarrow 0,\ \delta_k \downarrow 0$ as $k \to \infty$), a set of directions $Dir = \{(\check{p}^r, \check{z}^r, \check{x}^r, \check{v}^r) \in I\!R^{3nM+m} \mid (\check{p}^r, \check{z}^r, \check{x}^r, \check{v}^r) \neq 0,\ r = 1, ..., \mathcal{N}\}$, numbers $\xi_- \triangleq \inf(g, S)$ and $\xi_+ \triangleq \sup(g, S)$, and the algorithm parameters $\mathcal{M} \in \mathbb{N}$ and $\nu \in I\!R_+$.

Step 0. Set $k := 1$, $(\bar{p}^k, \bar{z}^k, \bar{y}^k, \bar{x}^k, \bar{v}^k) := (p^0, z^0, y^0, x^0, v^0)$, $r := 1$, $\xi := \xi_-$, $\triangle\xi = (\xi_+ - \xi_-)/\mathcal{M}$.

Step 1. Beginning from the point $(\bar{p}^k, \bar{z}^k, \bar{y}^k, \bar{x}^k, \bar{v}^k)$ and using the above local search procedure, construct a τ_k-critical point $(p^k, z^k, y^k, x^k, v^k) \in S$ in the problem $(\mathcal{P}(\sigma))$. Set $\zeta_k := \Phi(p^k, z^k, y^k, x^k, v^k)$.

Step 2. Using the 4-tuple $(\check{p}^r, \check{z}^r, \check{x}^r, \check{v}^r) \in Dir$ build the point $(\widetilde{p}^r, \widetilde{z}^r, \widetilde{x}^r, \widetilde{v}^r)$ from the approximation $\mathcal{A}_k = \{(\widetilde{p}^r, \widetilde{z}^r, \widetilde{x}^r, \widetilde{v}^r) \mid h(\widetilde{p}^r, \widetilde{z}^r, \widetilde{x}^r, \widetilde{v}^r) = \xi - \zeta_k,\ r = 1, ..., \mathcal{N}\}$ of the level surface $\mathcal{U}(\zeta_k) = \{(p, z, x, v) \mid h(p, z, x, v) = \xi - \zeta_k\}$ of the convex function $h(\cdot)$. Compute $\widetilde{y}^r$, where each coordinate $\widetilde{y}^r_{jt} := D_{jt} - \sum_{i \in I} \widetilde{x}^r_{ijt} (j \in J,\ t \in \Omega)$.

Step 3. If $g(\widetilde{p}^r, \widetilde{z}^r, \widetilde{y}^r, \widetilde{x}^r, \widetilde{v}^r) > \xi + \nu\xi$, $r < \mathcal{N}$ and $\xi < \xi_+$, then set $r := r+1$ and move to **Step 2**.

Step 4. If $g(\widetilde{p}^r, \widetilde{z}^r, \widetilde{y}^r, \widetilde{x}^r, \widetilde{v}^r) > \xi + \nu\xi$, $r = \mathcal{N}$ and $\xi < \xi_+$, then set $r := 1$, $\xi := \xi + \Delta\xi$ and go to **Step 2**.

Step 5. If $g(\widetilde{p}^r, \widetilde{z}^r, \widetilde{y}^r, \widetilde{x}^r, \widetilde{v}^r) > \xi + \nu\xi$, $r = \mathcal{N}$ and $\xi = \xi_+$, then stop; (p^k, y^k, x^k, v^k) is the obtained solution to the problem $(\mathcal{P}(\sigma))$.

Step 6. Find an approximate δ_k-solution $(\bar{p}^r, \bar{z}^r, \bar{y}^r, \bar{x}^r, \bar{v}^r)$ to the linearized problem $(\mathcal{PL}(\widetilde{p}^r, \widetilde{z}^r, \widetilde{x}^r, \widetilde{v}^r))$.

Step 7. Beginning from the point $(\bar{p}^r, \bar{z}^r, \bar{y}^r, \bar{x}^r, \bar{v}^r)$, construct the τ_k-critical point $(\hat{p}^r, \hat{z}^r, \hat{y}^r, \hat{x}^r, \hat{v}^r) \in S$ in the problem $(\mathcal{P}(\sigma))$ using the local search.

Step 8. If $\Phi(\hat{p}^r, \hat{z}^r, \hat{y}^r, \hat{x}^r, \hat{v}^r) < \zeta_k$, then set $\xi := \xi_-$, $(\bar{p}^{k+1}, \bar{z}^{k+1}, \bar{y}^{k+1}, \bar{x}^{k+1}, \bar{v}^{k+1}) := (\hat{p}^r, \hat{z}^r, \hat{y}^r, \hat{x}^r, \hat{v}^r)$, $k := k+1$, $r := 1$ and move to **Step 2**.

Step 9. If $\Phi(\hat{p}^r, \hat{z}^r, \hat{y}^r, \hat{x}^r, \hat{v}^r) \geq \zeta_k$, $r < \mathcal{N}$, then increase $r := r+1$ and go to **Step 2**.

Step 10. If $\Phi(\hat{p}^r, \hat{z}^r, \hat{y}^r, \hat{x}^r, \hat{v}^r) \geq \zeta_k$, $r = \mathcal{N}$ and $\xi < \xi_+$, then set $\xi := \xi + \triangle\xi$, $r := 1$ and return to **Step 2**.

Step 11. If $\Phi(\hat{p}^r, \hat{z}^r, \hat{y}^r, \hat{x}^r, \hat{v}^r) \geq \zeta_k$, $r = \mathcal{N}$ and $\xi = \xi_+$, then stop. $(p^k, z^k, y^k, x^k, v^k)$ is the obtained solution to the problem $(\mathcal{P}(\sigma))$. #

Recall that the algorithm parameters $\nu \in I\!R_+$ and $\mathcal{M} \in \mathbb{N}$ are responsible for minimizing rounding errors and dividing the interval $[\xi_-, \xi_+]$ into an appropriate number of subintervals (for the one-dimensional search along ξ), respectively. In this regard, it is important to note that the Global Search Scheme does not follow a traditional algorithmic structure, since some of its steps are not fully specified. For example, the selection of algorithm parameters and the construction of an approximation of the level surface of the function $h(\cdot)$ are not explicitly defined. Resolving these uncertainties depends on the available numerical data and is based on previous computational experiences in solving nonconvex problems [7,10,13–16]. As a result, final decisions on these aspects should be confirmed and justified by successful computational experiments.

6 Concluding Remarks

In this paper, we presented a new bilevel model for the economic system with one importer at the upper level and several exporters at the lower level. For this model, we developed new methods to find the global optimistic solution. These methods are grounded in the specialized Global Search Theory for nonconvex (DC) optimization proposed by A.S. Strekalovsky.

The primary target of future research in this area is collecting real data for the model (or at least plausible data for constructing test cases for the studied bilevel problem), as well as the implementation and evaluating the computational efficacy of the local and global search techniques. Based on the existing numerical experience (demonstrated by results in solving other bilevel problems [10–15]), we anticipate that the developed methodology is promising for efficiently solving the proposed bilevel model.

Acknowledgement. The research was funded by the Russian Science Foundation No. 24-41-03004, https://rscf.ru/project/24-41-03004/

References

1. Coal 2023—Analysis - IEA. International Energy Agency. https://www.iea.org/reports/coal-2023. Accessed 03 Jul 2025
2. Coal 2024—Analysis - IEA. International Energy Agency. https://www.iea.org/reports/coal-2024. Accessed 03 Jul 2025
3. Sxcoal—Premier Intelligence on China's Coal Market. https://www.sxcoal.com/en. Accessed 03 Jul 2025
4. Dempe, S., Zemkoho, A. (eds.): Bilevel Optimization: Advances and Next Challenges. Springer, Cham (2020)
5. Dempe, S.: Foundations of Bilevel Programming. Kluwer Academic Publishers, Dordrecht (2002)
6. Lim, S., Choi, K.: Sequential tariffs in Cournot and Bertrand competition. J. Econ. Theor. Econ. **35**(3), 73–96 (2024). https://doi.org/10.22812/jetem.2024.35.3.004 https://doi.org/10.22812/jetem.2024.35.3.004 https://doi.org/10.22812/jetem.2024.35.3.004
7. Strekalovsky, A.S.: Elements of Nonconvex Optimization. Nauka, Novosibirsk (2003). (in Russian)
8. Strekalovsky, A.S.: Global optimality conditions and exact penalization. Optim. Lett. **13**(3), 597–615 (2017). https://doi.org/10.1007/s11590-017-1214-x
9. Strekalovsky, A.S.: Minimizing sequences in a constrained DC optimization problem. Proc. Steklov Inst. Math. **323**(Suppl. 1), S255–S278 (2023). https://doi.org/10.1134/S0081543823060214
10. Strekalovsky, A.S., Orlov, A.V.: Linear and Quadratic-Linear Problems of Bilevel Optimization. SB RAS Publishing, Novosibirsk (2019). (in Russian)
11. Gruzdeva, T.V., Petrova, E.G.: Numerical solution of a linear bilevel problem. Comput. Math. Math. Phys. **50**(10), 1631–1641 (2010)
12. Strekalovsky, A.S., Orlov, A.V., Malyshev, A.V.: On computational search for optimistic solutions in bilevel problems. J. Global Optim. **48**, 159–172 (2010)

13. Orlov, A.V.: The global search theory approach to the bilevel pricing problem in telecommunication networks. In: Kalyagin, V.A., et al. (eds.) Computational Aspects and Applications in Large Scale Networks, pp. 57–73, Springer (2018). https://doi.org/10.1007/978-3-319-96247-4_5
14. Strekalovsky, A.S., Orlov, A.V.: Global search for bilevel optimization with quadratic data. In: Dempe, S., Zemkoho A. (eds.) Bilevel Optimization: Advances and Next Challenges, pp. 313–334, Springer, Cham (2020). https://doi.org/10.1007/978-3-030-52119-6_11
15. Orlov A.V., Minarchenko I.M., Gruzdeva T.V.: On a bilevel optimization model of electric power systems with a system operator at the upper level. In: Kochetov, Yu., et al. (eds.) Mathematical Optimization Theory and Operations Research, MOTOR 2025. LNCS, vol. 15681, pp. 286–301. Springer, Cham (2025). https://doi.org/10.1007/978-3-031-97077-1_20
16. Orlov, A.V., Strekalovsky, A.S., Batbileg, S.: On computational search for Nash equilibrium in hexamatrix games. Optim. Lett. **10**(2), 369–381 (2014). https://doi.org/10.1007/s11590-014-0833-8
17. Nocedal, J., Wright, S.J.: Numerical Optimization. Springer, New York (2006)

A Data-Driven Segmentation of Mongolia's Business Environment Using Machine Learning

Delgermaa Erkhemjargal[1], Bayanjargal Darkhijav[2(✉)], Davaasuren Batsukh[3], Sharbandi Ryenchin[3], and Batbaatar Chuluunbaatar[3]

[1] Institute of Mathematics and Digital Technology, Mongolian Academy of Sciences, Ulan Bator, Mongolia

[2] School of Information Technology and Electronics of Applied Mathematics, National University of Mongolia, Ulan Bator, Mongolia
bayanjargal@num.edu.mn

[3] Business School, National University of Mongolia, Ulan Bator, Mongolia

Abstract. The quality, accessibility, and stability of the business environment directly affect the competitiveness and growth of enterprises. Therefore, this study aims to classify Mongolia's business environment using machine learning techniques and identify each cluster's specific characteristics and key influencing factors. The research utilizes data from 4,073 enterprises operating in Mongolia as of 2020, comprising 104 variables related to the business environment. Since the original dataset contained many missing values, imputation was necessary. Mode and K-Nearest Neighbors (KNN) imputation methods—commonly used in research—were tested. Their mean absolute errors (MAE) were evaluated against the original data, and KNN (MAE = 0.0119) demonstrated higher accuracy than Mode (MAE = 0.0438). As a result, the KNN-imputed dataset was used for further analysis. Five clustering algorithms were applied to group the data, and their performance was assessed using three metrics: Silhouette Score, Davies-Bouldin Index, and Calinski-Harabasz Score. Among these, the HDBSCAN model yielded the best results (Silhouette Score = 0.435, DB Index = 0.409), and after removing 1,051 enterprises classified as noise, the remaining 2,852 enterprises were grouped into three clusters. Additionally, Random Forest classification and SHAP analysis were used to identify the most influential features in cluster formation. The study showed that bank interest rates, the effectiveness of government policies, and household living standards had the most significant impact on cluster differentiation. By identifying these key drivers, this study provides scientifically grounded insights that support evidence-based decision-making for policymakers and researchers seeking to improve the business environment.

Keywords: Clustering · Business Environment · Enterprise · KNN imputation · Random Forest

A. Chinchuluun et al. (Eds.): AAIM 2025, LNCS 16379, pp. 125–139, 2026.
https://doi.org/10.1007/978-981-95-5657-1_11

1 Introduction

The quality, accessibility, and stability of the business environment are critical determinants of firm competitiveness, investment attraction, and sustainable growth [1,2]. In developing countries, evaluating and segmenting business environments is key in informing policymaking and investment decisions [3]. In Mongolia, the "Mongolia Business Environment" national survey has been conducted every three years since 2017, providing structured insights for economic development policy [4].

However, conventional descriptive statistical methods—such as simple means, frequencies, and cross-tabulations—may not fully capture hidden subgroup structures or complex variable interactions within firm-level business environment data. This limitation has been increasingly recognized in international research, prompting machine learning and data-driven analytical methods to perform more nuanced segmentation and evaluate business environments.

One crucial preprocessing step in machine learning for survey data is handling missing values. Among widely applied techniques, Mode Imputation and K-Nearest Neighbors (KNN) Imputation are frequently used in business, social, and health sciences [5,6]. Mode Imputation offers simplicity but may distort distributions in skewed or complex datasets. At the same time, KNN Imputation leverages multivariate similarity and is shown to provide more accurate estimates in various business and economics studies [7].

Similarly, for clustering, numerous international studies have applied K-Means, Spectral Clustering, Gaussian Mixture Models (GMM), DBSCAN, and HDBSCAN to segment business-related, social, and regional datasets effectively. For instance, [8] successfully applied K-Means and GMM to market segmentation in manufacturing industries. [9] demonstrated Spectral Clustering's strengths in high-dimensional social science datasets. In urban and economic geography, [10] and [11] highlighted the robustness of DBSCAN and HDBSCAN when examining heterogeneous spatial patterns and detecting outliers in smart-city and entrepreneurial contexts, respectively.

DBSCAN and HDBSCAN have been increasingly favored for their ability to detect clusters of varying density and automatically identify outliers, as applied in smart city, economic geography, and entrepreneurship research [10,12].

Building on these international methodologies, this study applies a hybrid approach combining KNN imputation for missing data handling and HDBSCAN clustering to segment Mongolian enterprises based on 2020 business environment survey data. The objective is to identify structural differences among firms, uncover hidden subgroups, and determine key variables driving business environment quality.

2 Data and Methodology

2.1 Dataset

The 2020 Business Environment Survey of Mongolia is used, which includes respons-es from 4,073 formally registered enterprises. The dataset is structured to encompass 104 variables across five thematic domains:

- **Economic (E)**: Access to finance, costs, pricing, banking conditions
- **Governance (G)**: Legal trust, transparency, efficiency of government decision-making
- **Infrastructure (I)**: Access to utilities, transport, internet, and logistics
- **Social (S)**: Labor quality, social services, household living conditions
- **Organizational (O)**: Internal management, planning, leadership, and innovation

Each domain contains several subdimensions, reflected in the variable coding structure. For example, variables labeled as `E_1.*` refer to *taxation and pricing*, `E_2.*` to *financial markets and credit access*, and `E_3.*` to *labor and employment conditions*. Similarly, subdimensions are as follows:

`G_1.*`: Government institutional structure and coordination `G_2.*`: Business legal and regulatory environment `G_3.*`: Government performance and effectiveness.

`I_1.*`: Core physical infrastructure (utilities, roads, logistics) `I_2.*`: Technology and communications.

`S_1.*`: Social attitudes and trust `S_2.*`: Demographic and living conditions.

`O_1.*`: Organizational leadership and internal coordination `O_2.*`: Organizational productivity and operational efficiency.

This coding structure allows for both domain-level and subdomain-level analysis throughout the study.

2.2 Missing Value Imputation

The dataset contained missing values in multiple variables. We applied two commonly used imputation methods—Mode Imputation and K-Nearest Neighbors (KNN) to ensure analytical integrity. Mode imputation replaces missing values with the most frequent value, whereas KNN imputation estimates missing values using the closest k neighbors in Euclidean space [13].

To compare the accuracy of these methods, we calculated the Mean Absolute Error (MAE) against a subset with complete entries.

Missing values are shared in survey-based datasets, especially in business environment assessments [14]. Two widely adopted imputation methods—Mode Imputation and KNearest Neighbors (KNN) Imputation—were applied and compared in this study.

Mode Imputation. Mode Imputation replaces missing values with the most frequently occurring value in the same variable (feature). This simple and computationally efficient approach makes it suitable for categorical and ordinal data [15].

Mathematically, for a variable $X = \{x_1, x_2, \ldots, x_n\}$, the imputed value x_i for a missing observation is defined as:

$$x_i = \arg\max_{v \in V} P(X = v)$$

where V is the set of observed unique values in X, and $P(X = v)$ is the empirical probability of the value v.

K-Nearest Neighbors (KNN) Imputation. KNN Imputation estimates missing values based on the values of the k most similar records [6]. Similarity is typically measured using Euclidean distance for continuous variables or Gower distance for mixed data types.

Let $X = \{x_{i1}, x_{i2}, \ldots, x_{ip}\}$ be a p-dimensional feature vector with missing entries. The imputed value $\hat{x}_{ij}$ is computed as follows:

$$\hat{x}_{ij} = \frac{1}{k} \sum_{l \in \mathcal{N}_k(i)} x_{lj} \tag{1}$$

Where: x_{lj} are the observed values of feature j in the k nearest neighbors,
$\mathcal{N}_k(i)$ denotes the set of k nearest neighbors to instance i, considering only non-missing features.

The performance of each imputation method was assessed using two standard error metrics: Mean Absolute Error (MAE) and Root Mean Squared Error (RMSE). These metrics provided quantitative measures of accuracy by comparing imputed values against observed data, enabling an objective comparison between the methods.

2.3 Clustering Techniques

This study evaluated five clustering algorithms commonly used in machine learning and data science: K-Means, DBSCAN, HDBSCAN, Spectral Clustering, and Gaussian Mixture Models (GMM). Each method exhibits unique theoretical foundations and practical implications.

K-Means Clustering: K-Means is a partition-based clustering method introduced by [16]. It minimizes intra-cluster variance while maximizing inter-cluster separation.

Mathematically, K-Means seeks to minimize the following objective function:

$$J = \sum_{i=1}^{k} \sum_{x \in C_i} \|x - \mu_i\|^2 \tag{2}$$

Where: C_i is the set of points in the cluster i, and
μ_i is the centroid of the cluster i.

DBSCAN (Density-Based Spatial Clustering of Applications with Noise): DBSCAN, introduced by [17], defines clusters as high-density regions separated by low-density areas.

HDBSCAN (Hierarchical DBSCAN): HDBSCAN, proposed by [18], extends DBSCAN by building a hierarchy of density-based clusters and condensing it into a flat clustering.

Spectral Clustering: Spectral Clustering uses the spectrum (eigenvalues) of a similarity matrix derived from the input data to reduce dimensionality before clustering [9].

Gaussian Mixture Models (GMM): GMMs model data as a mixture of multiple Gaussian distributions [19]. Clustering is performed using the Expectation-Maximization (EM) algorithm.

Probability density function:

$$p(x) = \sum_{i=1}^{k} \pi_i \mathcal{N}(x \mid \mu_i, \Sigma_i) \tag{3}$$

Where: π_i are the mixing coefficients,
$\mathcal{N}(x \mid \mu_i, \Sigma_i)$ is the Gaussian density.

To evaluate and compare clustering model performance, three standard internal validation metrics were applied: Silhouette Score [20], Davies–Bouldin Index [21], and Calinski–Harabasz Index [22]. These metrics provide complementary insights into cluster compactness, separation, and structure.

2.4 Feature Importance

We applied Random Forest classification alongside SHAP (Shapley Additive exPlanations) value analysis to identify which variables contributed most significantly to cluster formation. Random Forest ranks variable importance by measuring how much each feature reduces classification impurity across all decision trees [23]. SHAP values provide a complementary perspective by attributing contribution scores to each feature for individual predictions based on cooperative game theory, offering global and local interpretability [24].

In this study, Random Forest was trained using cluster labels derived from HDBSCAN on KNN-imputed business environment data. Feature importances were determined via Gini impurity reduction, while SHAP values were computed using TreeExplainer to confirm and interpret results.

3 Results

A more detailed, data-driven classification and assessment of Mongolia's business environment is essential for policymakers and researchers to understand structural differences and enterprise-level heterogeneity better. Traditional statistical methods often fail to capture hidden structures and latent subgroups within firm-level data. Therefore, this study applies modern machine learning techniques—including K-Nearest Neighbors (KNN) imputation, HDBSCAN clustering, and Random Forest analysis—to perform a more granular segmentation of the business environment and identify key differentiating variables. The following section presents the main findings derived from these methods.

The following graph illustrates the trade-off between the allowed percentage of missing values (NA threshold) and the number of enterprises retained in the dataset (see Fig. 1).

As the threshold increases from 40% to 80%, the number of retained enterprises rises from approximately 3,650 to over 4,050. Based on this analysis, a 60% NA threshold was chosen, maintaining 3,905 enterprises (approximately 96% of the original sample), ensuring data representativeness while minimizing missingness.

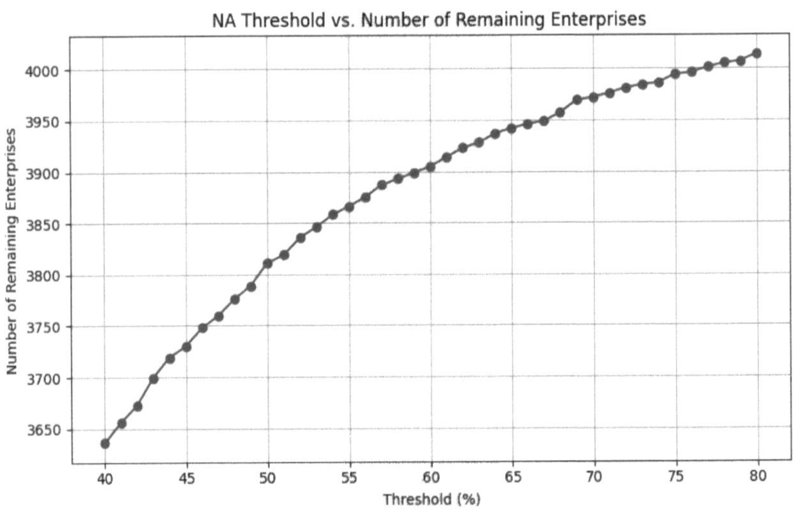

Fig. 1. Relationship Between NA Threshold and Number of Remaining Enterprises.

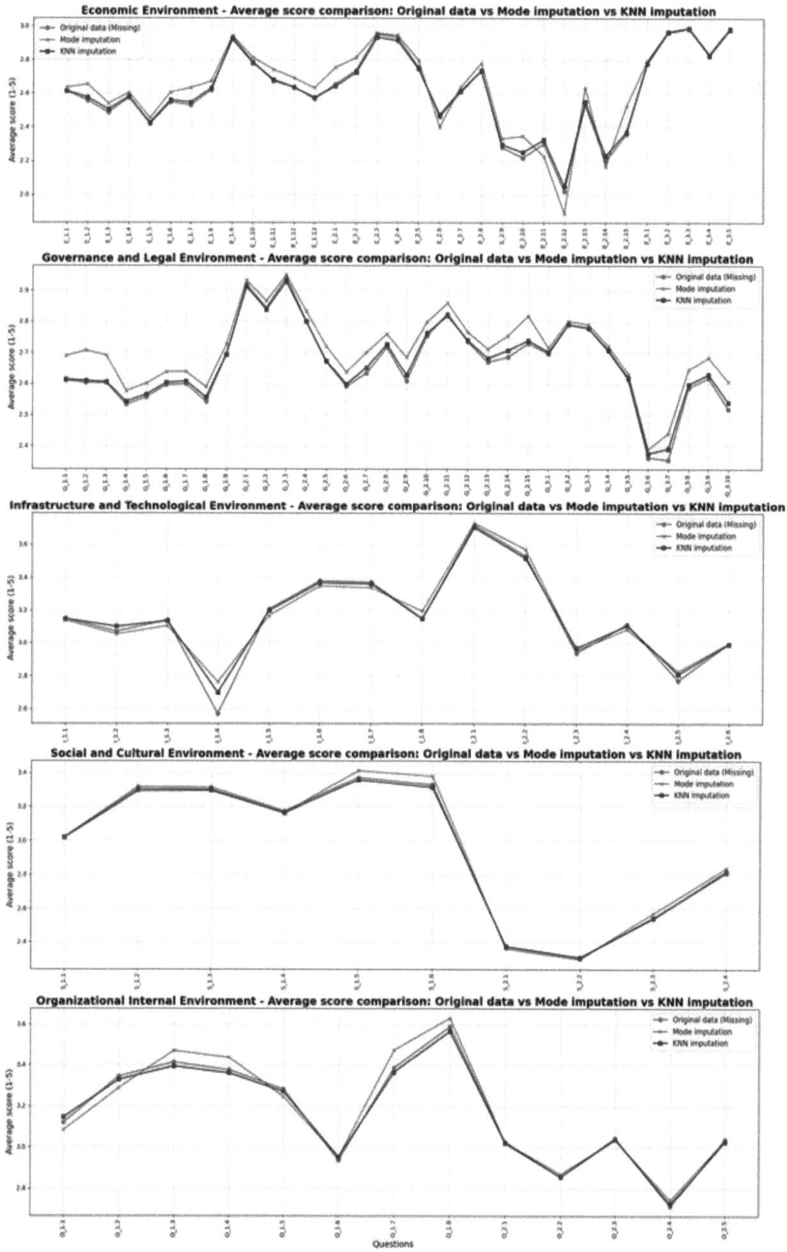

Fig. 2. Comparison of Original, Mode Imputation, and KNN Imputation Across Five Business Environment Domains.

(See Fig. 2) This set of five-line graphs compares the average scores by question under three conditions: original data with missing values, Mode Imputation, and KNN Imputation. The results show that KNN Imputation provides scores closer to the original data pattern, especially in domains with continuous or highly varied values, such as Economic and Governance. Mode Imputation tended to flatten variability, particularly in Organizational Environment scores.

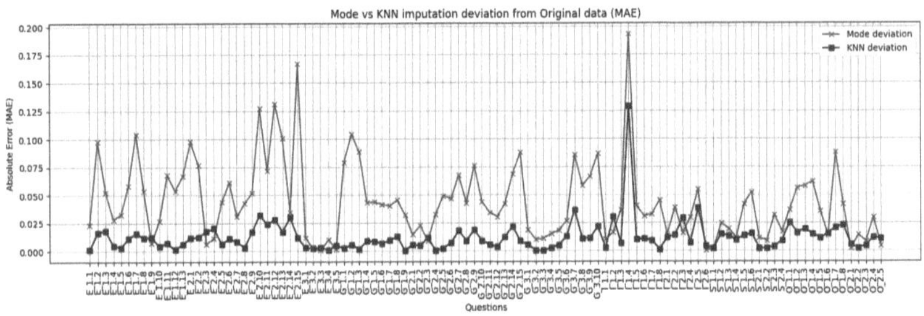

Fig. 3. Mean Absolute Error (MAE) Between Imputed and Original Values Across All Variables.

Before applying clustering algorithms, Principal Component Analysis (PCA) was performed on both Mode and KNN imputed datasets to reduce dimensionality and visualize the structural separability of enterprises. This transformation into two principal components allowed for clearer observation of potential groupings in the data and facilitated more robust clustering.

(See Fig. 3) shows MAE values for each variable, comparing the Mode and KNN imputation methods. KNN consistently achieved lower MAE across most variables. A notable outlier is I_1.4 (Sea access), where both methods show larger deviation. This variable's outlier status is attributed to Mongolia's landlocked geography, making sea access structurally irrelevant for many firms. Overall, KNN imputation demonstrated superior accuracy for handling missing values in this business environment dataset.

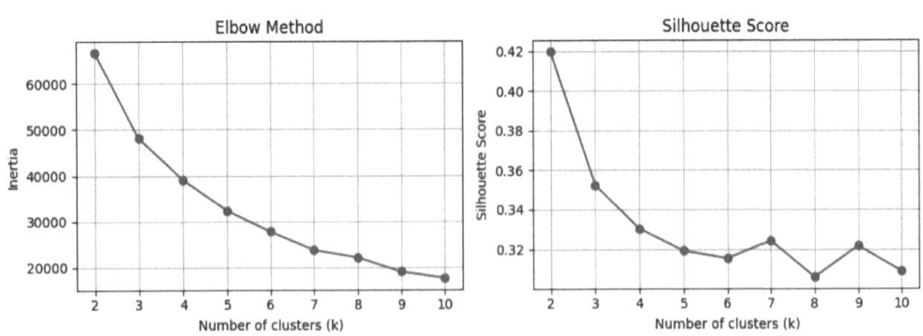

Fig. 4. Elbow Method and Silhouette Score for Cluster Number Selection.

These two plots summarize the model selection process for clustering (See Fig. 4). The Elbow Method graph shows inertia decreasing with the number of clusters, leveling off after $k = 3–4$. The Silhouette Score peaks at $k = 2$ but remains reasonably stable for $k = 3–4$. Balancing statistical validity with business interpretability, $k = 3$ was selected as the final number of clusters for analysis.

Table 1. Comparison of Clustering Models.

#	Model	Silhouette Score (↑)	Davies–Bouldin Index (↓)	Calinski–Harabasz Score (↑)
1	HDBSCAN	0.44	0.41	168.45
2	Gaussian MixtureModels	0.37	0.86	2997.45
3	K-Means	0.37	0.90	3447.30
4	Spectral Clustering	0.36	0.91	3416.80
5	DBSCAN	0.05	0.61	57.66

Table 1 presents the comparative performance evaluation of five clustering algorithms applied to the business environment dataset: HDBSCAN, Gaussian Mixture Models, K-Means, Spectral Clustering, and DBSCAN. Performance was assessed using three key internal validation metrics:

- Silhouette Score (higher is better) - HDBSCAN achieved the highest silhouette score of 0.44, indicating better-defined and well-separated clusters.
- Davies–Bouldin Index (lower is better) - HDBSCAN again showed the lowest Davies-Bouldin Index at 0.41, suggesting compact and distinct clusters.
- Calinski–Harabasz Score (higher is better) - While K-Means (3447.30) and Spectral Clustering (3416.80) scored higher on this metric, they performed worse on the other two indicators.

Considering all three metrics collectively, HDBSCAN was identified as the most balanced and robust clustering method for segmenting Mongolia's business environment dataset.

The horizontal bar chart ranks the five most influential organizational variables (See Fig. 5). "Organizational structure and internal coordination" (O_1.2) and "Leadership capacity and effectiveness" (O_1.5) stand out as the highest-impact variables. These factors play a critical role in distinguishing business environment quality across clusters, highlighting areas for policy and organizational improvement.

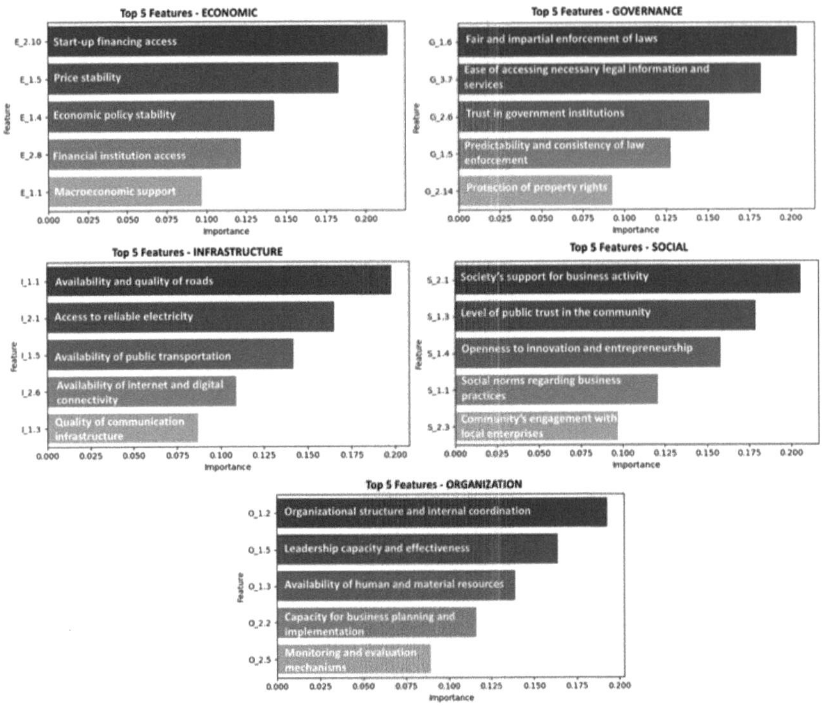

Fig. 5. Top 5 most important features across five business environment domains.

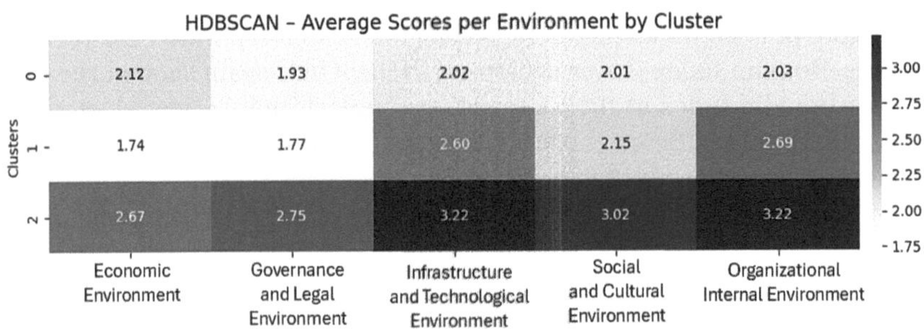

Fig. 6. Cluster-Wise Average Scores by Business Environment Domain.

The heatmap visualizes average scores across the five business environment domains for the three identified clusters (See Fig. 6). Cluster 2 demonstrates the highest scores in all dimensions, particularly Infrastructure and Organizational Internal Environment (average ≈3.22). Cluster 0 shows the lowest scores, especially in Economic (1.74) and Governance (1.77), confirming its designation as a "Challenging Environment." Cluster 1 represents a transitional profile with moderate scores. This distribution validates the clustering structure and helps interpret group-level business conditions.

Table 2. Business Duration Distribution Across Clusters.

Cluster	10 years or more	6–10 years	1–5 years	1 year
0	17%	26%	44%	13%
1	42%	29%	29%	0%
2	37%	30%	29%	5%

Table 2 presents the distribution of business duration across the three identified clusters. Cluster 0 consists mainly of younger firms, with 44% operating for only 1–5 years. Cluster 1 and Cluster 2 show larger shares of firms with over 10 years of operation, at 42% and 37% respectively. This indicates structural differences in business maturity levels among clusters.

Table 3. Employee Size Distribution Across Clusters.

Cluster	Up to 10	11–50	51–200	200 or more
0	74%	16%	11%	0%
1	57%	44%	0%	0%
2	61%	26%	8%	4%

Table 3 shows the number of employees per enterprise within each cluster. Cluster 0 is dominated by micro-enterprises with up to 10 employees (74%). Cluster 1 has a higher share of small to medium-sized firms, with 44% having 11–50 employees. Cluster 2 includes a small portion (4.2%) of large enterprises with over 200 employees.

Table 4. Annual Revenue Distribution Across Clusters.

Cluster	Up to 50M MNT	51–300M MNT	301M–1B MNT	1B MNT or more
0	59%	18%	9%	14%
1	39%	35%	17%	9%
2	41%	30%	14%	15%

Table 4 summarizes the annual revenue distribution among clusters. Cluster 0 primarily comprises lower-revenue firms, with 59% earning up to 50 million MNT annually. Cluster 1 and Cluster 2 have larger proportions of firms earning higher revenues, with Cluster 2 showing the highest share (15%) of firms reporting annual revenues exceeding 1 billion MNT.

Average score differences across the five business environment domains were calculated to quantify the improvements needed for enterprises in underdeveloped environments (Cluster 0) to reach conditions like those in favorable environments (Cluster 2).

(See Fig. 7) As shown in the figure, the Infrastructure and Technological Environment and the Organizational Internal Environment require the most substantial improvements, with each needing an average increase of approximately 1.2 score units (around 59%). The Governance and Legal Environment also shows a considerable gap, requiring an improvement of 0.82 units (~42%). In contrast, the Economic Environment exhibits the smallest gap between clusters, requiring only a 0.55-unit improvement (~26%).

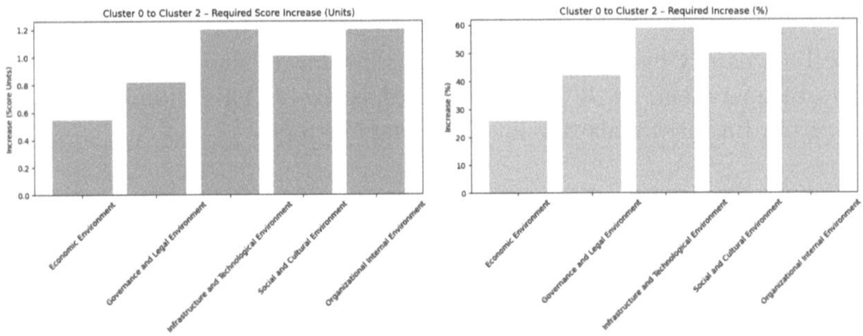

Fig. 7. Business Environment Improvement Requirements.

4 Discussion

This study demonstrated the utility of a machine learning-based segmentation approach to classify and analyze the Mongolian business environment. Applying KNN imputation and HDBSCAN clustering to the 2020 Business Environment Survey dataset identified three distinct enterprise clusters, each exhibiting specific structural and performance characteristics.

Cluster 0 reflected firms facing challenging conditions, with notably low scores in the economic, governance, infrastructure, and organizational domains. These businesses were predominantly small, young and had lower revenue and fewer employees. Cluster 1 represented a transitional profile, showing moderate scores in the domains. Although these companies demonstrated better organizational and infrastructure conditions than Cluster 0, weaknesses in governance and economic factors remained evident. Cluster 2 consisted of firms operating in a favorable environment, marked by high average scores in all five domains, especially in infrastructure and organizational capacity. Cluster 2 firms were generally older, larger, and financially stronger, as evidenced by higher revenue and employment figures. The most significant disparities across clusters were observed in infrastructure and organizational capacity, requiring over 1.2 score units of improvement for Cluster 0 firms to match Cluster 2 levels. This finding underscores the critical role of these two domains in shaping enterprise-level business environment quality.

Furthermore, the analysis of the importance of the Random Forest features revealed that variables such as organizational structure and internal coordination (O_1.2), leadership capacity (O_1.5), loan interest rates (E_1.5), income distribution (E_1.4), and effectiveness of policy implementation (G_1.6) were the most influential factors differentiating clusters. These insights align with previous international research on business environment segmentation and validate the application of KNN imputation and HDBSCAN clustering as effective methods for handling missing data and detecting complex group structures in economic datasets.

Although the analysis identified loan interest rates, governance effectiveness, and internal organizational capacity as key differentiating factors between groups, more comparative reflection is warranted to contextualize these findings. Prior studies in emerging economies such as Vietnam, Indonesia, and Georgia have similarly emphasized the importance of governance and credit access in shaping firm performance and perceptions of the business environment [25–27]. In Mongolia's context, the salience of loan interest rates (E_1.5) may be heightened by structural barriers in financial markets and limited access to affordable credit—especially among small and medium-sized enterprises (SMEs). Likewise, governance-related challenges such as regulatory ambiguity and enforcement gaps may amplify external constraints for firms. These elements may reflect broader trends in developing economies but also point to context-specific vulnerabilities. Future research could benefit from a cross-country cluster comparison to further examine which patterns are generalizable and which are uniquely shaped by Mongolia's institutional and economic setting.

While this study is grounded in the 2020 Business Environment Survey, it is important to acknowledge its temporal limitations. At the time this research was conducted, the detailed microdata from the 2023 Business Environment Survey had not yet been made publicly available. Nevertheless, we reviewed and incorporated key preliminary findings from the 2023 summary report published by the Mongolian National Chamber of Commerce and Industry (MNCCI) [28]. These findings indicate a general decline in the overall business environment ratings compared to 2020, particularly in governance, legal, infrastructure, and socio-cultural domains. In contrast, a slight improvement was observed in the Internal Organizational Environment, potentially reflecting firms' adaptive responses during the post-pandemic recovery period. These emerging trends appear broadly consistent with the structural patterns revealed by our cluster analysis—particularly the disparities between internal and external domains. Applying the proposed segmentation framework to the complete 2023 microdata, once released, would be a valuable direction for future research. Such analysis would enable a more robust assessment of cluster stability and provide deeper insight into evolving business conditions in Mongolia.

5 Conclusion

This study demonstrated that Mongolia's business environment can be effectively segmented into three distinct groups-Challenging, Transitional, and Favorable-using a machine learning-based approach combining KNN imputation and HDBSCAN clustering. The results highlight that enterprises in Cluster 0, representing the most challenging environment, require the most significant improvements, particularly in infrastructure and organizational capacity, with an average gap of over 1.2 score units compared to Cluster 2. Internal organizational factors, such as structure, leadership, and resource availability, were key differentiators shaping business environment quality across clusters. While the economic domain also plays a role, it exhibited comparatively minor disparities. Based on these findings, the study emphasizes the need for cluster-specific policy interventions focused on infrastructure, governance, and organizational development, primarily targeting enterprises in Clusters 0 and 1. Future research should apply this model to newer datasets to track changes over time and refine the segmentation approach.

References

1. World Bank: Doing Business 2020: Comparing Business Regulation in 190 Economies. World Bank, Washington, DC (2020). https://doi.org/10.1596/978-1-4648-1440-2
2. OECD: SME and Entrepreneurship Policy in Mongolia 2021: Assessment and Recommendations. OECD Publishing, Paris (2021). https://www.oecd.org/en/publications/2021/12/oecd-economic-outlook-volume-2021-issue-2-b3069fed.html
3. UNDP: Inclusive Business Environment Report 2022. United Nations Development Programme, New York (2022). https://www.undp.org/publications/undp-annual-report-2022
4. Mongolian National Chamber of Commerce and Industry (MNCCI): Mongolia Business Environment Survey 2020. Ulaanbaatar (2020)
5. Azur, M.J., Stuart, E.A., Frangakis, C., Leaf, P.J.: Multiple imputation by chained equations: what is it and how does it work? Int. J. Meth. Psychiatr. Res. **20**(1), 40–49 (2011). https://doi.org/10.1002/mpr.329
6. Batista, G.E.A.P.A., Monard, M.C.: An analysis of four missing data treatment methods for supervised learning. Appl. Artif. Intell. **17**(5–6), 519–533 (2003). https://doi.org/10.1080/713827181
7. Yang, Q., Zhao, Z., Chen, M.: Missing data imputation methods in business big data analytics: a review. Big Data Res. **25**, 100230 (2021). https://doi.org/10.1016/j.bdr.2021.100230
8. Xu, D., Tian, Y.: A comprehensive survey of clustering algorithms. Ann. Data Sci. **2**(2), 165–193 (2015). https://doi.org/10.1007/s40745-015-0040-1
9. Luxburg, U.V.: A tutorial on spectral clustering. Stat. Comput. **17**(4), 395–416 (2007). https://doi.org/10.1007/s11222-007-9033-z
10. Campello, R.J.G.B., Moulavi, D., Sander, J.: Hierarchical density estimates for data clustering, visualization, and outlier detection. ACM Trans. Knowl. Discov. Data **10**(1), 5 (2015). https://doi.org/10.1145/2733381
11. McInnes, L., Healy, J., Astels, S.: HDBSCAN: hierarchical density-based clustering. J. Open Source Softw. 2(11), 205 (2017). https://doi.org/10.21105/joss.00205
12. Elhamifar, E., Vidal, R.: Sparse subspace clustering: algorithm, theory, and applications. IEEE Trans. Pattern Anal. Mach. Intell. **35**(11), 2765–2781 (2013)

13. Troyanskaya, O., Cantor, M., Sherlock, G., et al.: Missing value estimation methods for DNA microarrays. Bioinformatics **17**(6), 520–525 (2001). https://doi.org/10.1093/bioinformatics/17.6.520
14. Little, R.J.A., Rubin, D.B.: Statistical Analysis with Missing Data, 3rd edn. Wiley, Hoboken (2019)
15. Acuna, E., Rodriguez, C.: The treatment of missing values and its effect on classifier accuracy. In: Classification, Clustering, and Data Mining Applications, pp. 639–647. Springer, Berlin (2004). https://doi.org/10.1007/978-3-642-17103-1_60
16. MacQueen, J.B.: Some methods for classification and analysis of multivariate observations. In: Proceedings of the 5th Berkeley Symposium on Mathematical Statistics and Probability, vol. 1, pp. 281–297. University of California Press, Berkeley (1967). http://projecteuclid.org/euclid.bsmsp/1200512992
17. Ester, M., Kriegel, H.P., Sander, J., Xu, X.: A density-based algorithm for discovering clusters in large spatial databases with noise. In: KDD, vol. 96, pp. 226–231 (1996)
18. Campello, R.J.G.B., Moulavi, D., Sander, J.: Density-based clustering based on hierarchical density estimates. In: Advances in Knowledge Discovery and Data Mining, pp. 160–172. Springer, Berlin (2013)
19. McLachlan, G., Peel, D.: Finite Mixture Models. Wiley Series in Probability and Statistics. Wiley, Hoboken (2000). https://doi.org/10.1002/0471721182
20. Rousseeuw, P.J.: Silhouettes: a graphical aid to the interpretation and validation of cluster analysis. J. Comput. Appl. Math. **20**, 53–65 (1987). https://doi.org/10.1016/0377-0427(87)90125-7
21. Davies, D.L., Bouldin, D.W.: A cluster separation measure. IEEE Trans. Pattern Anal. Mach. Intell. **1**(2), 224–227 (1979). https://doi.org/10.1109/TPAMI.1979.4766909
22. Calinski, T., Harabasz, J.: A dendrite method for cluster analysis. Commun. Stat. **3**(1), 1–27 (1974). https://doi.org/10.1080/03610927408827101
23. Breiman, L.: Random forests. Mach. Learn. **45**(1), 5–32 (2001). https://doi.org/10.1023/A:1010933404324
24. Lundberg, S.M., Lee, S.-I.: A unified approach to interpreting model predictions. In: Proceedings of the 31st International Conference on Neural Information Processing Systems, NeurIPS 2017, Long Beach, USA, 4–9 December 2017, pp. 4766–4777 (2017). https://proceedings.neurips.cc/paper_files/paper/2017/file/8a20a8621978632d76c43dfd28b67767-Paper.pdf
25. World Bank: Bank Lending Rates and Spreads in Emerging Markets and Developing Economies (EMDEs). World Bank Open Knowledge Repository (2023). https://openknowledge.worldbank.org/entities/publication/7ada0b63-44f1-55ea-a24b-e53e73d90421
26. International Finance Corporation (IFC): ESG performance and loan contracting in an emerging market. Pacific-Basin Finan. J. **78** (2023). https://doi.org/10.1016/j.pacfin.2023.101987
27. International Finance Corporation (IFC): Governance and Performance in Emerging Markets. IFC Publications, Washington, DC (2019). https://www.ifc.org
28. Mongolian National Chamber of Commerce and Industry (MNCCI): Preliminary Summary of the 2023 Business Environment Survey. Ulaanbaatar (2023). https://mongolchamber.mn/a/1489

Simulating Call Center Operations: A Case Study of Ulaanbaatar City's Public Transportation

Dolgorsuren Gombojav[1] and Mend-Amar Majig[2(✉)]

[1] Graduate School of Arts and Letters, Tohoku University, Sendai, Japan
[2] School of Information Technology and Electronics, National University of Mongolia, Ulan Bator, Mongolia
mendamarm@num.edu.mn

Abstract. In this work, we studied the operational effectiveness of the Call Center of the Public Transport Department (PTD) of Ulaanbaatar. Our focus was on its responsiveness to incoming calls, review of the workload based on the actual data, and the integration of gender and social inclusion issues. This research enables the PTD to make evidence-based decisions regarding human resources while also addressing critical gender and social inclusion concerns. Enhancing the performance and efficiency of the PTD is crucial for improving public transportation services, a significant contribution to alleviating the pressing issue of traffic congestion in Ulaanbaatar.

We conducted statistical and simulation analyses using queueing theory on historical data from March 22 to September 20, 2023. Our findings indicate that the call center receives 12 calls on average, spending 12.4 min each hour. However, 31% (7,185 calls during weekdays) went unanswered. Our analysis helps distinguish whether missed calls result from line congestion or were simply ignored, providing valuable insights into call center performance.

Additionally, we found that the PTD's current classification of complaints and requests lacks the depth required to fully understand and address the diverse experiences of public transport users.

Keywords: data analysis · queuing theory · simulation · public transport · call center

1 Introduction

The Public Transport Department (PTD) of Ulaanbaatar, the capital city of Mongolia, bears a significant responsibility toward the city's residents and the Government of Mongolia. Originally planned for a population of 400,000, Ulaanbaatar now holds 1.7 million residents, nearly half of Mongolia's total population. This overpopulation has significantly contributed to traffic congestion, highlighting PTD's challenges. One of the most effective strategies to tackle

A. Chinchuluun et al. (Eds.): AAIM 2025, LNCS 16379, pp. 140–157, 2026.
https://doi.org/10.1007/978-981-95-5657-1_12

traffic congestion is making public transport an attractive alternative to private vehicle use [24]. Operating under public scrutiny, the PTD strives to improve the existing bus transport system, running 1,043 buses across 106 routes as of December 2023.

The PTD recognizes the importance of public feedback for providing responsive services and attracting more customers. The PTD communicates with the public transport users in a variety of ways and listens to their travel experiences and voices, and integrates them into its service planning. One key communication channel is the call center hotline, 700440xx, which operates under the Inspection Division. The call center handles complaints and requests from the public, operating daily from 7 AM to 10 PM. Even though the PTD states that it aims to respond to every call, there is a common belief among the public that the call center staff may not be making sufficient efforts to handle all incoming calls. This points to the need for a thorough evaluation of the existing data. One of the efforts made by the PTD was increasing the number of call center staff from one to two in 2023, though they have yet to assess the impact of this change on responsiveness. Without analyzing data on the response rate of incoming calls and staff performance, it remains difficult to determine the effectiveness of these adjustments and to clarify whether dropped calls are due to call congestion or other factors.

Improving the effectiveness of the call center is essential not only for evaluating staff performance but also for making informed decisions about resource allocation [1]. Analyzing call records by problem type is critical, as most complaints received at the PTD relate to accessibility issues, such as long waiting times, overcrowding and inconsistent services, and lack of real-time information along with reckless driving and driver misconduct. Age, disability, and different extents traveling with young children or with heavy luggage are the main factors that constrain people's mobility. Thus, it is the transport should address the problems faced by the most vulnerable in the society [17].

However, the current complaint records lack necessary disaggregation by demographic information and location, which hinders the PTD's ability to take targeted actions where needed. Addressing this gap by disaggregating data will not only improve public transport services but also help in tackling social inclusion challenges. Transport-related inclusion is a complex interplay of factors, where accessibility plays a vital role in determining whether a citizen can fully participate in socio-economic life [33]. The historical records of the complaints and requests received at the call center indicate many face accessibility issues. Without disaggregated data, the PTD cannot effectively address the specific needs of different demographic groups, limiting the inclusivity and effectiveness of its services.

Our analysis first examined call center performance regarding responsiveness to incoming calls and evaluated the actual workload and staff required for efficient operation. Additionally, we reviewed the complaint system's integration of gender and social inclusion issues in terms of disaggregation of callers' demographic information. We conclude that effective staffing and detailed analysis of complaints are crucial for the PTD to operate more efficiently, address recurring problems, and better serve all segments of Ulaanbaatar's population.

The rest of this paper is organized as follows: First, "Literature Review" explores existing studies on call center operations, queuing theory, and the importance of addressing social inclusion in public transportation. Next, "Data Analysis and Insights" presents the findings from historical records and synthetic data generation, forming the foundation of our queueing model. The "Statistical Analysis" subsection explains the approach, with a special focus on the simulation techniques applied to assess call center performance and workload distribution. The simulation allows us to model real-world scenarios, providing valuable insights into how the call center operates under various conditions and enabling data-driven decision-making for resource allocation. The "Simulation Analysis for Call Center Process" section discusses the results, compares them to the optimal workload defined by call center operators, and provides policy implications and recommendations. Finally, "Conclusion" summarizes the key findings and offers suggestions for future research and improvements in public transportation services.

2 Literature Review

Call centers are the most preferred means of communication for many organizations with their customers [21]. A pool of resources, including agents and information communication technology equipment, makes it possible to handle customer contacts by telephone. Call centers, both in the public and private sectors, share a goal of obtaining at least a certain minimal service level for minimal costs. It is increasingly used in public organizations to fulfill the needs of the public instead of increasing their competitiveness in the open market [3,32]. Satisfying the public needs, the public transport services can provide quality services with regular schedules, safe and easy journeys with the optimal mobilization of the sources [7,20,25].

The provision of public transport has been globally recognized to be one of the most effective ways to reduce congestion (Buchanan, 2019). However, the quality of the services, public satisfaction, and perception of public transport determine the preference for public transport over the use of private vehicles [31]. Often it becomes the call centers' responsibility to collect data about the quality of public transport service provided and sense the public satisfaction with it. Various scientific fields from mathematics and statistics, through operations research, industrial engineering, information technology, and human resource management, all the way to psychology and sociology come into play to investigate the complexity of the system and effectiveness of the call center operation and help authorities to make evidence-based strategic decision, which does not solely rely on workforce management [21].

The service level of a call center is determined by a number of criteria, one of which is the responsiveness of the call center, including waiting time to reach a call center agent along with the quality of the service provided by telephone

[26]. Thus, businesses need to design sophisticated systems over two competing goals of meeting the goals of responding rapidly and avoiding overstaffing as it, in turn, increases the business expense [36]. However, keeping labor expenses as low as possible while keeping the service quality as high as possible has always been one of the top priorities for managers [8]. To measure the effectiveness and efficiency of call centers and determine an optimal number of staff and their working hours, the number of calls received and the total of time spent on responding to calls and after-call work should be taken into account [12]. To determine an appropriate staffing requirement function $s(t)$, which defines the number of servers to be assigned at each time instant t (with $0 \leq t \leq T$), the analysis should be conducted on the historical data.

Before that, a desired level of responsiveness or system performance, namely waiting time, should be decided as customer satisfaction with the service is directly related to the waiting times and abandonment of calls [22] suggested the following three operational regimes, including efficiency driven regime, quality-driven regime, and rationalized regime. By definition, an efficiency-driven regime places higher importance on keeping labor costs as minimal as possible than maintaining service quality, which often results in a high probability of delay and loss of calls. A quality-driven regime, on the other hand, may lead to high staffing costs because of excess staff. Therefore, the rationalized or quality-and-efficiency driven (QED) regime promises the most optimal solution to this problem as it helps to provide good service while keeping staffing costs acceptably low [15].

In the context of public transport, accessibility refers to the ease with which all individuals, regardless of their physical, economic, or social status, can access and use transportation services. Accessibility is a critical factor that influences the overall inclusiveness of public transport systems. A transport system with high accessibility ensures that people from all walks of life—especially those who are elderly, disabled, or economically disadvantaged—can participate fully in social, economic, and cultural activities [28]. This accessibility is not just a matter of physical access but also includes affordability, availability, and the ease of use of the transport services [19].

Accessibility is intrinsically linked to social inclusion, as it determines whether individuals can fully engage in society. Social inclusion in public transport means that transport services are designed and operated in a way that considers the diverse needs of the population, including marginalized and vulnerable groups [27]. Gender issues are particularly pertinent in this context, as women often face unique challenges in accessing public transport. These challenges may include safety concerns, the availability of transport options during times of day when women are more likely to travel, and the design of transport services that accommodate those with caregiving responsibilities [23].

Addressing social inclusion and gender issues in public transport requires a comprehensive approach that involves collecting disaggregated data on how different groups experience transport services [34]. Call centers can play a crucial role in this process by gathering detailed feedback from users, which can then be analyzed to identify barriers to access and areas where services can be improved.

This data-driven approach allows public transport authorities to tailor services to meet the specific needs of all users, thereby promoting greater social inclusion [5,6]. The effectiveness of call centers in public transport is not only a matter of operational efficiency but also a critical component of ensuring that public transport systems are accessible and inclusive for all members of society. By integrating considerations of social inclusion and gender into call center operations and decision-making processes, public transport authorities can enhance the overall quality and equity of their services [13].

3 Disaggregating Public Transport Complaints for Social Inclusion

As shown in Tables 1 and 2, the existing categories for complaints and requests within the PTD provide only a general overview but fail to disaggregate data by crucial demographic factors such as age, gender, disability status, and location. Moreover, the categories are broad and often overlap, for instance, "poor driver behavior" and "using mobile phones while driving" both broadly relate to safety concerns. This lack of specificity and granularity in the data prevents the PTD from effectively addressing accessibility or social inclusion issues that particularly affect people with disabilities, elderly passengers, or those traveling from more remote areas of the city.

Table 1 details several critical areas of concern within the PTD's operations. The highest number of complaints is attributed to 'Poor driver behavior,' accounting for 30.6% of the total, indicating significant issues with how drivers interact with passengers or manage their driving responsibilities. This is followed by complaints related to 'Violation of traffic rules, speeding' at 6.3%, and 'Using mobile phones while driving' at 5.6%, highlighting ongoing safety risks within the public transport system. Additionally, operational failures such as buses failing to stop at designated stops, which make up 3.13% of the complaints, directly compromise the reliability and accessibility of the service.

In contrast, the requests primarily highlight the public's demand for infrastructure improvements. All actionable requests during this period were concentrated on the need to 'Open or organize bus stops and routes,' representing 100% of such requests. This concentration indicates significant deficiencies in the current route planning and stop accessibility, which may not adequately meet the needs across all urban areas, particularly in less centrally located regions.

The data from these tables demonstrate that the PTD's current categorization system lacks the depth required to fully understand and address the diverse experiences of public transport users. This is particularly true in terms of accessibility and social inclusion, where more detailed, disaggregated data would allow the PTD to identify and rectify specific challenges faced by different demographic groups.

To enhance the inclusivity of the public transport system, it is crucial to refine the categorization of complaints and requests to include more detailed demographic factors such as age, gender, disability status, and location. Such

Table 1. Summary of Complaints and Responses

No.	Type of Complaints	Resolved and Responded		Overdue		Total	
		N	%	N	%	N	%
1	Bus did not stop at designated stops	5	3.13	0	0	5	3.13
2	Bus did not announce the stop	0	0	0	0	0	0
3	Poor driver behavior	49	30.6	0	0	49	30.6
4	Violation of traffic rules, speeding	10	6.3	0	0	10	6.3
5	Route deviation	2	1.3	0	0	2	1.3
6	Non-compliance with standards	1	0.6	0	0	1	0.6
7	Smoking on the bus	23	14.4	0	0	23	14.4
8	Using mobile phones while driving	9	5.6	0	0	9	5.6
9	Failure to pick up passengers	13	8.1	0	0	13	8.1
10	Non-compliance with bus schedules	1	0.63	0	0	1	0.63
11	Issues related to bus cleanliness and appearance	0	0	0	0	0	0
12	Poor behavior of ticket inspectors	0	0	0	0	0	0
13	Service-related issues with bus routes	0	0	0	0	0	0
14	Bus not operating	1	0.63	0	0	1	0.63
15	Miscellaneous service-related issues	42	26.3	0	0	42	26.3
16	Issues related to fare and pricing discrepancies	4	2.5	0	0	4	2.5
Total		160	100	0	0	160	100

Table 2. Summary of Requests and Responses

No.	Type of Requests	Resolved and Responded		Overdue		Total	
		N	%	N	%	N	%
1	Increase the number of buses	0	0	0	0	0	0
2	Improve the appearance and cleanliness of buses	0	0	0	0	0	0
3	Open or organize bus stops and routes	2	33.33	0	0	2	33.33
4	Issues related to UB Smart Card payment system	0	0	0	0	0	0
5	Issues related to law and standards	0	0	0	0	0	0
6	Reduce turnaround times and organize departure points	0	0	0	0	0	0
7	Information and training related to services and drivers	0	0	0	0	0	0
8	Compliance with service schedules and timetables	0	0	0	0	0	0
Total		2	33.33	0	0	2	33.33

disaggregation is essential as it can reveal that certain unsafe or discriminatory practices might be more prevalent in less monitored or economically disadvantaged areas.

The issues identified suggest that the PTD needs to take effective measures to make the existing service more accessible and inclusive. Implementing systems for collecting more granular data on the demographics and locations of complainants will help tailor services and address specific community needs. Focusing on behavioral training for drivers, especially in areas with higher rates of complaints, will ensure safety and professional conduct across all city areas. Additionally, a thorough review of current bus infrastructure based on the requests for more organized bus stops and routes will improve accessibility, especially in underserved areas.

4 Data Analysis and Insights

The data analysis stage involves several steps, from data cleaning and preliminary data processing to generating synthetic data for in-depth analysis based on pre-determined criteria. In terms of the sources of the data, we used the PTD call center records for six months and qualitative data obtained through workplace observation and interviews with the call center operators. The absence of demographic data, such as age, gender, and location, limits the ability to conduct a fully disaggregated analysis, particularly in addressing social inclusion challenges. This gap highlights the need for future improvements in data collection processes. Using this information, we employed bootstrapping techniques to generate synthetic data that simulated the call arrival patterns, allowing for further analysis under different operational conditions (see Table 3).

Table 3. Headings of the Raw Data

Date	CallerID	Disposition	Who Hangup	Agent	Hold Time
2023-03-22 07:14:11	919120xx	ABANDONED			80
2023-03-22 07:16:28	997767xx	TIMEOUT			120
2023-03-22 07:19:48	919120xx	ANSWERED	CALLED	129	114
2023-03-22 07:23:40	951117xx	ABANDONED			78
...	...	...	...	...	...
2023-09-19 21:52:43	967108xx	ANSWERED	CALLED	500	12
2023-09-19 21:55:37	995077xx	ANSWERED	CALLED	500	5
2023-09-19 22:14:42	808192xx	ANSWERED	CALLED	500	11
2023-09-20 08:20:40	883816xx	ABANDONED			5

Quantitative the PTD provided was 31,090 call records registered between March 22 and September 20, 2023. Prior to conducting the actual data analysis, the dataset was cleaned according to specific criteria to define what qualifies as a valid call. First, only calls received during working hours (7 AM to 10 PM) were retained, reducing the dataset to 29,141 calls. Next, the dataset was filtered to include only weekdays, further reducing the number to 23, 288 calls. The purpose of this research is to evaluate the call center process and its responsiveness and staff workload during regular business days. Since the average number of calls is significantly higher on weekdays (179 calls) compared to weekends (113 calls), focusing on the higher call volume allows for a more accurate assessment of staff workload during peak operational periods.

4.1 Statistical Analysis

Before going deep into analysis based on the queueing theory, we primarily conducted the descriptive analysis to produce descriptive statistics to get an insight into the workload of the call center and its operators. The descriptive analysis process helped us to come up with the primitives of the call center model, which include arrival rate (λ), the number of servers (N), and the average service time ($1/\mu$) [8].

To gain deeper insight into the existing workload of the call center operators, we conducted analysis by the total number of calls received over a six-month period and further examined the distribution of calls across working hours to determine the average number of calls handled per hour. As shown in Fig. 1, the call centers experience their peak workload at 10 AM and 9 PM, with 14.9 and 15.1 calls handled, respectively. The hourly average number of calls is 12, with each call lasting an average of 89.7 s, totaling 17.94 min per hour.

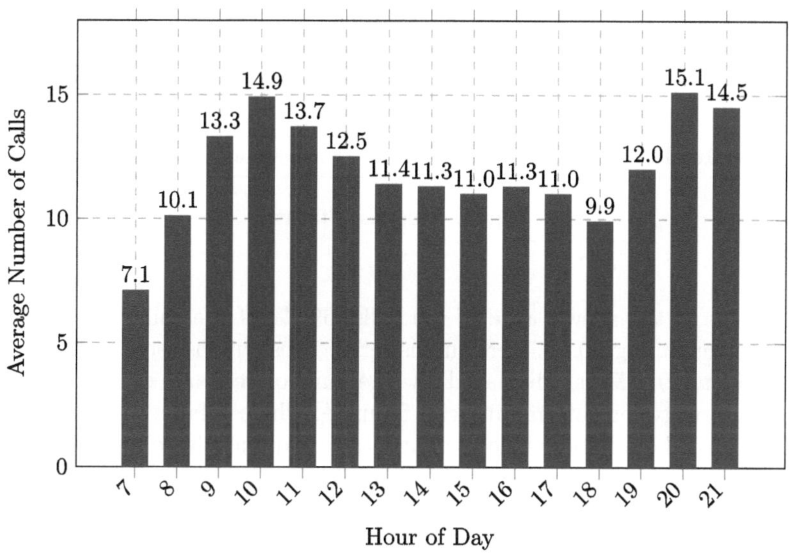

Fig. 1. Average Number of Hourly Calls.

Figure 2 illustrates the total number of calls received during working hours over the given period, along with the success of the call attempts. The graph clearly shows that most unanswered calls occur at the end of the day (around 9 PM) and during the peak hour (10 AM). To clarify whether these unanswered calls were due to call congestion or other factors, we calculated the end time of each call based on the sum of arrival time, hold time, and call duration. If the arrival time of an unanswered call does not fall within this timeframe, we consider that the call could have been answered, as the line was free at that time. Through this method, we identified the volume of ignored calls among all unanswered calls. The figure shows that most unanswered calls occurred during idle hours. Figure 3 presents the exact portion of ignored calls among all unanswered calls.

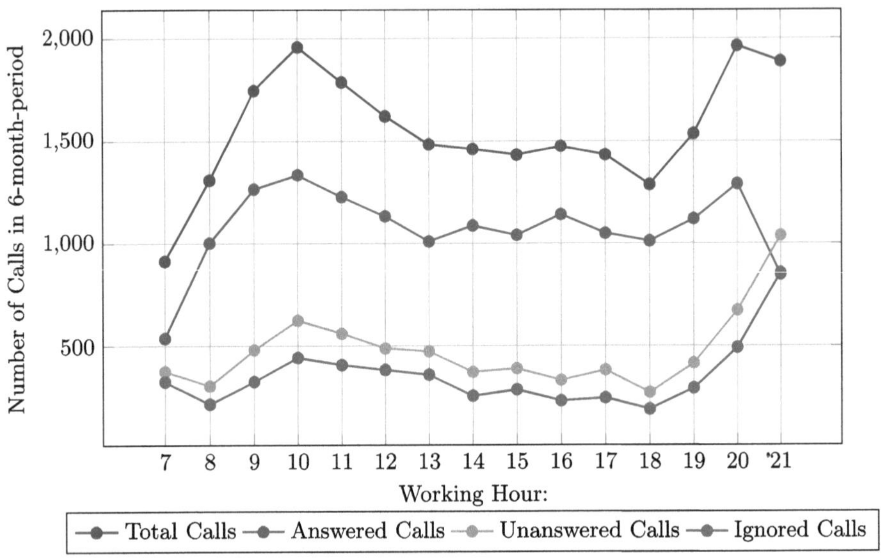

Fig. 2. Response Status.

In terms of the answer rate of the calls, 69.5% of the calls were answered, while the remaining 30.5% were unanswered. Most unanswered calls occurred at 7 AM and 9 PM, as displayed by the sum of red and orange. Based on our methodology for determining call failures—where a failure is defined as an unanswered call occurring while the line was free and available to take the call—around two-thirds of all unanswered calls could have been answered. Specifically, for the calls at 7 AM and 9 PM, 35.7% of the 41.2% unanswered calls at 7 AM happened when the line was free, and 45.3% of the 54.8% unanswered calls at 9 PM occurred under the same condition. We further analyzed the percentage of calls generated by the same caller within 30 min after the first attempt failed. Out of 7,190 unanswered calls, 3,900 (54.8%) were generated within 30 min simply because the initial attempt(s) failed. These calls can be considered artificially generated and should not be counted as unique complaints or requests. Figure 4 shows the number of calls coming from unique callers, where 33% of the calls were from the same individuals.

As mentioned in the previous section, the queueing theory was an integral part of our analysis as it allowed us to understand the process and estimate the workload of the staff. In other words, we incorporated the fundamentals of the queueing system into our analysis, including the arrival process, server capacity, queueing discipline, and departure process. Emerged during the 1900s in A.K.

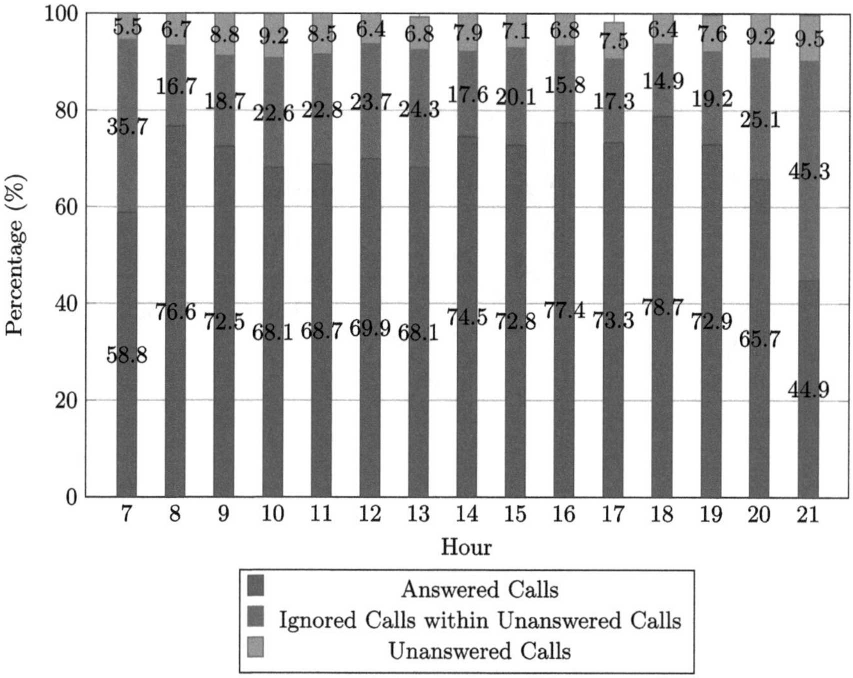

Fig. 3. Hourly Ignored Calls.

Erlang [18] work for the analysis tool and design of the telephone system, the queueing theory has evolved several times. Preceded by the renewed application of codification in the field of operation research in the 1950s, queueing theory started gaining popularity in the performance analysis of the time-shared computer system. In many kinds of literature, it is noted that much of the early work in queueing theory was inspired by problems regarding telephone traffic [29], and it can be applied to call center performance optimization [9]. One of the most used techniques to derive the queueing formula is to write the birth-and-death equation, which is translated into the arrival and departure process [14]. In order words, queueing theory explains the arrival and service process determined heavily by a service facility [16].

Arrival Process Analysis. The randomness of customer arrival often characterizes the uncertainty of the service availability, including that of call centers [30]. Despite the complexity of the system and external factors influencing the performance and characteristics of call centers, modeling is essential to determine the arrival process patterns and manage sources accordingly [4,35]. It is common that the arrival process of call centers is assumed to be a Poisson process, or stochastic, where calls arrive identical in terms of distribution and independently of each other within a given time frame [10].

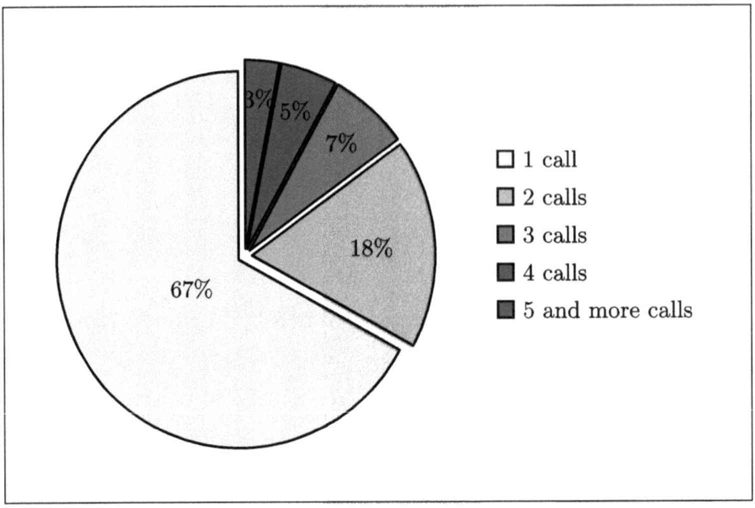

Fig. 4. Distribution of events by occurrence frequency.

Therefore, we assume our calls to be exponentially distributed and draw a histogram to visualize the distribution (See Fig. 5). Although the distribution of the calls appears to be exponential, we perform the Kolmogorov-Smirnov (KS) test to assess the goodness-of-fit, evaluating how well the observed distribution aligns with the expected exponential distribution (See Fig. 6). For this test, we generate synthetic data using the same rate parameter ($\lambda = 0.0353$) as the actual data. The Kolmogorov-Smirnov test yields a KS statistic of 0.0332 and a p-value of 1.61×10^{-6}. This result leads us to reject the null hypothesis, which assumes that the interarrival times of the calls are exponentially distributed.

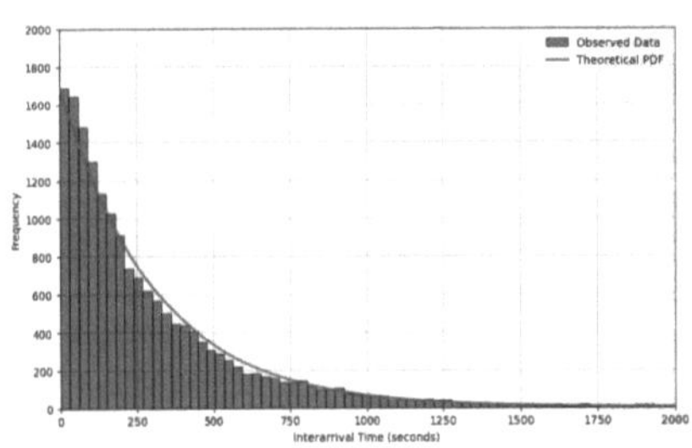

Fig. 5. Interarrival Time.

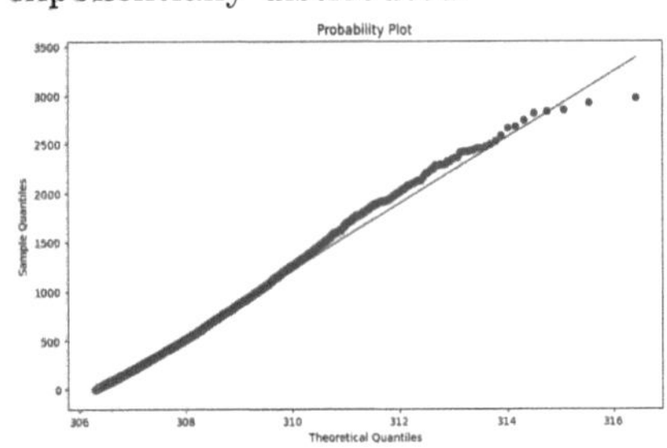

Fig. 6. Interarrival Time (KS Test).

Service Time Analysis. Most queueing theories assume service time follows an exponential distribution [2,11]. We applied the same process used for the arrival times: constructing a distribution graph, estimating the parameter of the service time data, and generating synthetic data with the same rate parameter

($\lambda = 0.0111$). The KS test yielded a KS statistic of 0.1495 and a p-value of 5.30×10^{-122}, leading to the rejection of the null hypothesis (H_0). This result indicates that the service time distribution differs from the theoretical exponential distribution (Fig. 8).

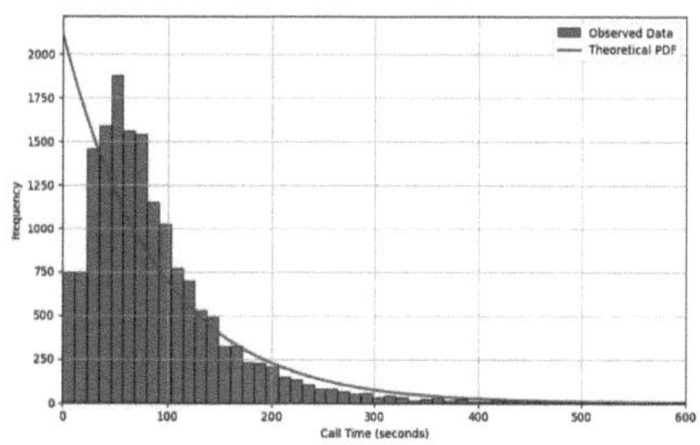

Fig. 7. Service Time.

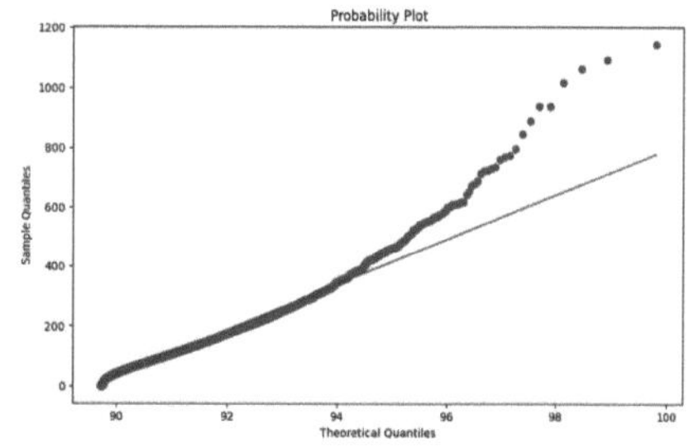

Fig. 8. Service Time (KS Test).

5 Simulation Analysis for Call Center Process

The operations of the PTD call center are straightforward. If an operator is available, they will answer and record the complaint when a call comes in. Otherwise, the caller is placed in a waiting queue until the operator completes the ongoing call. The maximum hold time in the queue is 120 s, after which the call is automatically disconnected. Some callers who fail to connect may attempt to call again later.

The team conducted a simulation modeling to assess the efficiency of the PTD call center's operations, focusing on how calls are handled, the efficiency of servers (operators), and the overall performance over a specified period. The simulation results provide insights into call arrival patterns by using random interarrival times and first call times from the past data. The simulation also provides queue and server management and performance metrics, including the number of calls answered or ignored in the given period, as well as returning callers, which adds realism to the simulation.

The general structure of our simulation model is depicted in Fig. 9. Incoming calls are placed in a first-in-first-out (FIFO) queue until they can be transferred to a hotline operator. The operators are modeled by the Server block, which may represent either a single server or multiple servers, depending on the number of operators. The Sink block signifies the completion of a call or caller abandonment.

The Call block generates calls based on an Interarrival time parameter derived from historical data using the bootstrapping technique (Fig. 5). This parameter determines the time interval before the next call arrives. Additionally, each call is assigned attributes for Hold time and Call time, also generated from historical data through bootstrapping (Fig. 7). The model incorporates a Resilience parameter, with historical data indicating a probability of 0.548, which reflects the likelihood of a caller attempting to reconnect after a failed attempt to reach an operator.

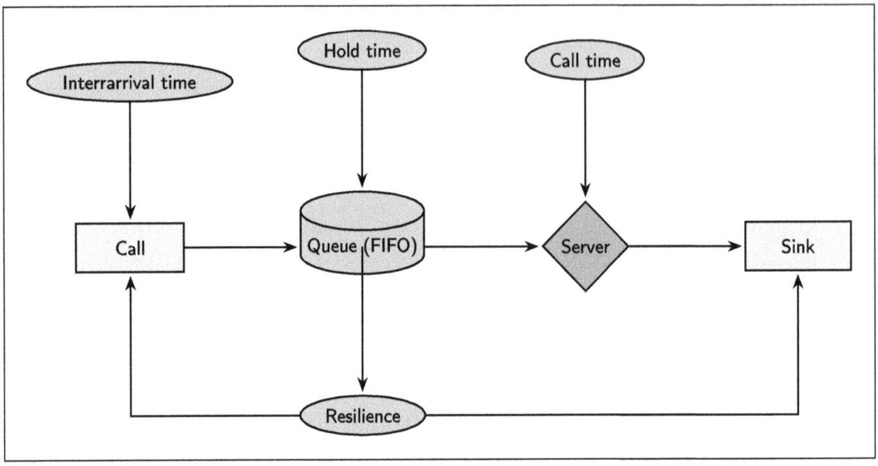

Fig. 9. Simulation Diagram for Call and Server Workflow.

We simulated our model over 130 days, representing a six-month period of weekdays. Throughout the simulation, we measured several key indicators, including the number of daily calls, hourly call volume, calls generated after failed attempts, server occupancy rates based on the number of servers, and the number of unanswered calls relative to the number of available operators.

Figure 10 shows the simulation results for the daily calls and the artificially generated calls—those created when the initial call attempt fails—depicted in blue and red, respectively. As illustrated in the figure, the average number of daily calls was 162.7, with 11.3 being artificially generated.

Figure 11 shows the simulation results of the hourly call volume over the given period. These results are compared to the optimal workload reported by call center operators, an average of 15 calls per server or operator. The red lines in the figure indicate the appropriate workload thresholds for one operator. While the average call volume per hour is 12.0, the simulation results indicate that the hourly call volume exceeds the optimal workload for 14.7% of the time. However, when a 2-h moving average is applied—allowing for workload fluctuations within that timeframe—this excess workload decreases to 9.5% of the time. Therefore, we can conclude that a single server is sufficient to handle incoming calls, as the workload remains below the optimal threshold set by the call center operators approximately 85% of the time.

In Fig. 12, we observe that when only one operator is on duty, the server is busy over 20% of the time, with an average occupancy rate of 23.3%. For the case with two operators, the occupancy rate resulted in an average occupancy rate of 12.4%, while with three operators working, the occupancy rate further

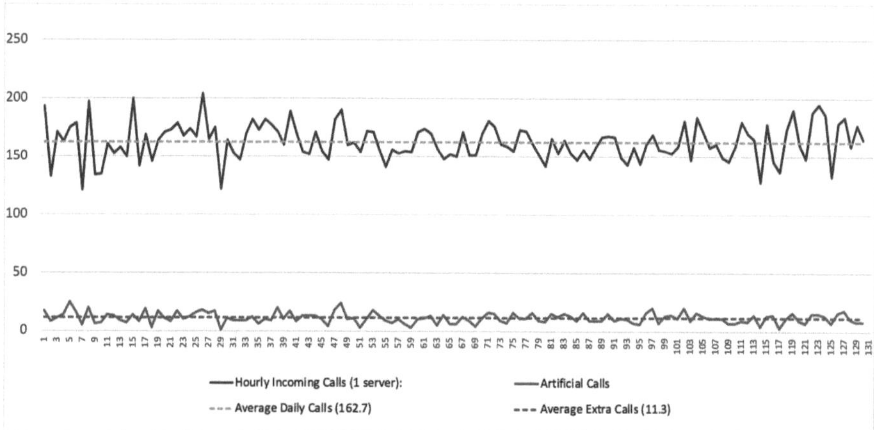

Fig. 10. Daily Call Simulation.

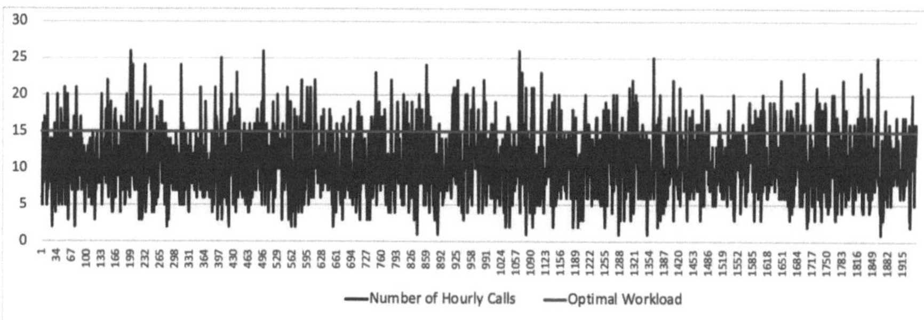

Fig. 11. Hourly Call Simulation.

declined to 8.3%. This data also indicates that having more than one operator is hardly necessary.

For customer comfort, we also evaluated the percentage of unanswered calls due to a busy line with one, two, and three operators on duty (see Fig. 13). When only one operator works, an average of 13.7% of calls go unanswered. However, with two operators, this percentage drops significantly to 1.9%.

In summary, the simulation results highlight the operational efficiency of the PTD call center under varying conditions. A single operator can manage the call load effectively for approximately 85% of the time, with call volume rarely exceeding the optimal threshold. However, during peak hours, adding a second operator significantly reduces the percentage of unanswered calls and lowers the server occupancy rate, improving overall responsiveness. While three operators further decrease occupancy and unanswered calls, the data suggests that employing more than two operators is unnecessary for regular operations. Therefore, strategic staffing adjustments during peak periods, such as using two operators,

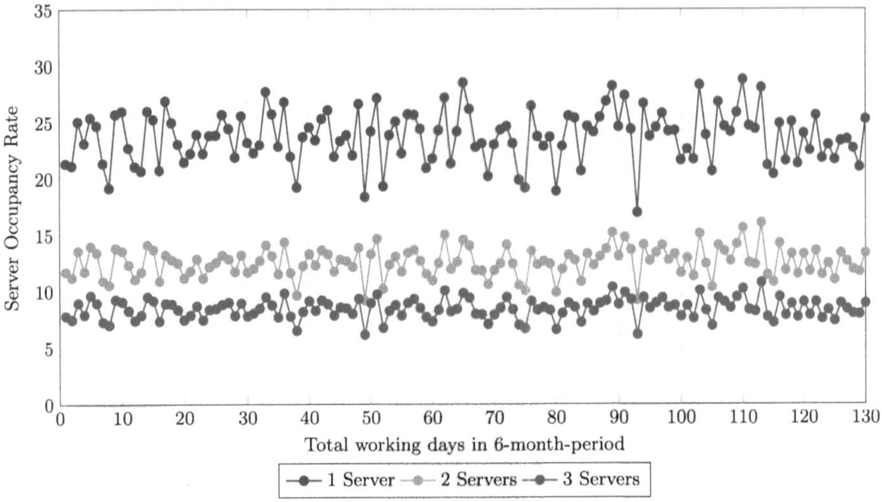

Fig. 12. Server Occupancy Rate.

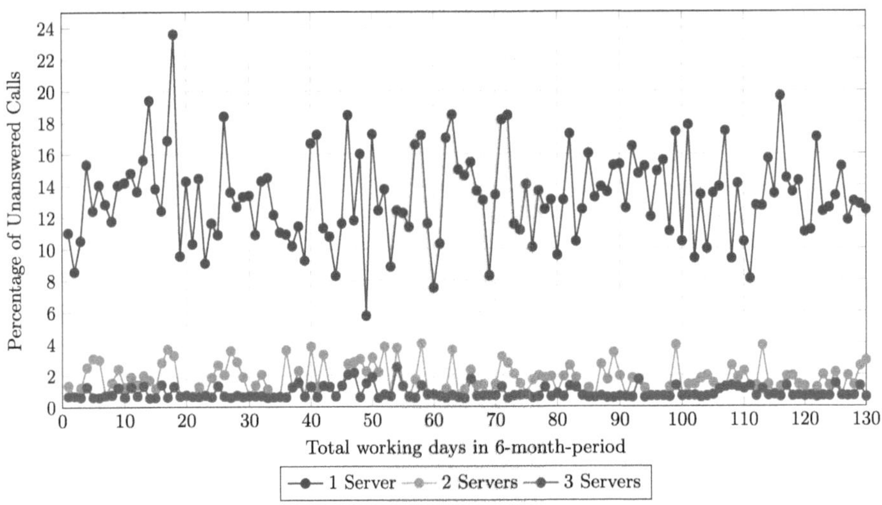

Fig. 13. Share of Unanswered Calls.

would optimize both efficiency and service quality while avoiding unnecessary resource expenditure.

6 Conclusion

This study analyzed the operational effectiveness of the PTD's call center in Ulaanbaatar, focusing on its call-handling efficiency, responsiveness to complaints, and the integration of gender and social inclusion concerns. Using queueing theory and simulation techniques, we conducted a detailed examination of call handling patterns, resource allocation, and performance metrics. The simulation modeled the arrival of calls, queuing behavior, and server capacity, providing valuable insights into how calls are managed and how staffing levels impact the overall efficiency of the call center.

Our findings indicate that while 69.5% of calls were answered, a significant portion (30.5%) went unanswered, with peak failures occurring at 7 AM and at 9 PM. The simulation results showed that the average hourly call volume exceeded the optimal workload by only 14.7% of the time, with the peak hour workloads reaching 15.1 calls, compared to the optimal rate of 15 calls per server. When applying a 2-h moving average, the excess call volume was reduced to 9.5%. Furthermore, the simulation revealed that one server could handle the incoming call volume approximately 85% of the time. However, during peak hours, the number of unanswered calls increased, particularly during early morning (7 AM) and late evening (9 PM) periods, where 45.3% of the unanswered calls occurred when the line was free.

These simulation results were instrumental in providing evidence-based recommendations for optimizing resource allocation. The simulation demonstrated how the call center could operate more efficiently with dynamic staffing during peak demand hours by modeling interarrival times, call durations, and hold times.

In addition to operational improvements, our analysis highlighted the need for the PTD to collect more disaggregated demographic data, including age, gender, and disability status. This would enable the PTD to address social inclusion issues and tailor services to better meet the needs of all users, particularly marginalized groups such as the elderly, women, and people with disabilities. Without this data, the PTD's ability to address specific complaints related to accessibility and inclusivity is limited. We recommend that the PTD implements a more detailed data collection system that disaggregates caller demographics to improve service delivery for all groups, particularly marginalized ones.

In conclusion, the combination of simulation techniques and data analysis is crucial for optimizing the call center's operations and enhancing the overall quality of public transport services in Ulaanbaatar. We recommend not only optimizing staffing during high-demand periods but also implementing a system to collect more detailed demographic information from callers to address social inclusion and improve service delivery.

Acknowledgement. This research was supported by the National University of Mongolia (P2023-4614) and the Mongolia–Japan Engineering Education Development Project (JR14B16).

References

1. Aksin, Z.A., Armony, M., Mehrotra, V.: The modern call center: a multi-disciplinary perspective on operations management research. Prod. Oper. Manage., 665–688 (2007)
2. Alsaadi, Z., Ibrahim, M.: Wait time estimation in distributed multitenant systems using queueing theory. Stockholm, Sweden (2022)
3. Anable, J.L.: 'Complacent car addicts' or 'aspiring environmentalists'? Identifying travel behaviour segments using attitude theory. Transp. Policy, 65–78 (2005)
4. Avramidis, A.N., Deslauriers, A., L'Ecuyer, P.: Modeling daily arrivals to a telephone call center. Manage. Sci., 896–908 (2004)
5. Bank, A.D.: Gender Tool Kit: Transport—Maximizing the Benefits of Improved Mobility for All. Asian Development Bank (2013)
6. Bank, T.W.: Progress Towards Gender Equality in the Middle East and North Africa Region. World Bank (2017)
7. Belwal, R., Belwal, S.: Public transportation services in Oman: a study of public perception. J. Public Transp. **13**(4) (2010)
8. Borst, S., Mandelbaum, A., Reiman, M.I.: Dimensioning large call centers. Oper. Res. **52**(1), 17–34 (2004)
9. Brezavšček, A., Baggia, A.: Optimization of a call centre performance using the stochastic queueing models. Bus. Syst. Res., 6–18 (2014)
10. Brown, L., Gans, N., Mandelbaum, A.: Statistical analysis of a telephone call center: a queueing-science perspective. J. Am. Stat. Assoc., 36–50 (2005)
11. Brown, L., Gans, N., Mandelbaum, A.: Statistical analysis of a telephone call center: a queueing-science perspective. J. Am. Stat. Assoc. (2005)
12. Chawla, A., Shah, S.: Improving quality and agent's performance in Bangkok call centres. J. Supply Chain Manage. Res. Pract. **13**(1) (2019)
13. European Commission: She Moves: Women's Issues in Transportation. European Commission (2014)
14. Cooper, R.B.: Queueing theory: a 90-minute tutorial. Florida Atlantic University (1981)
15. Defraeye, M., Van Nieuwenhuyse, I.: Setting staffing levels in an emergency department: opportunities and limitations of stationary queueing models. Rev. Bus. Econ., 73–100 (2011)
16. Dombacher, D.C.: Introduction to Queueing Theory. Wagram Austria (2010)
17. Organisation for Economic Co-operation and Development: Economic Benefits of Improving Transport Accessibility. OECD (2016)
18. Erlang, A.K.: The theory of probabilities and telephone conversations. Nyt. Tidsskr. Mat. Ser. B **20**, 33–39 (1909)
19. Farrington, J., Farrington, C.: Rural accessibility, social inclusion, and social justice: towards conceptualization. J. Transp. Geogr., 1–12 (2005)
20. Francis, L., Richard, J.: Customer service quality management in public transport: the case of rail transport in Tanzania. Int. Rev., 3–4 (2017)
21. Gans, N., Koole, G., Mandelbaum, A.: Telephone call centers: tutorial, review and research prospects. Manuf. Serv. Oper. Manage. **5**(2), 79–141 (2003)
22. Garnett, O., Mandelbaum, A., Reiman, M.: Designing a telephone call center with impatient customers. Manuf. Serv. Oper. Manage. **4**(3), 208–227 (2002)
23. Hanson, S.: Gender and mobility: new approaches for informing sustainability. Gender Place Culture, 5–23 (2010)

24. Kajita, D.T., et al.: Traffic congestion and impact on the environment in Vietnam: development of public transport system - experience from actual operation of bus in Hanoi. J. Civ. Environ. Eng. (2018)
25. Khalid, U.A., Lee, S.B.: User perceptions of rail public transport services in Kuala Lumpur, Malaysia: KTM Komuter. Procedia. Soc. Behav. Sci. **153**, 566–573 (2014)
26. Koole, G., Mandelbaum, A.: Queueing models of call centers: an introduction. Ann. Oper. Res. **41**(1), 41–59 (2002)
27. Litman, T.: Evaluating Accessibility for Transport Planning. Victoria Transport Policy Institute (2024)
28. Lucas, K.: Transport and social exclusion: where are we now? Transp. Policy, 105–113 (2012)
29. Newell, G.: Applications of Queueing Theory. Chapman and Hall Ltd. (1982)
30. Oreshkin, B.N., Reiman, M.I., Wang, W.: Rate-based daily arrival process models with application to call centers. Oper. Res., 510–527 (2016). INFORMS
31. de Oña, J., Eboli, L., Mazzulla, G.: Public transport users versus private vehicle users: differences about quality of service, satisfaction and attitudes toward public transport in Madrid (Spain). Travel Behav. Soc., 76–85 (2021)
32. Parsuraman, A., Zeithaml, V.A., Berry, L.L.: A conceptual model of service quality and its implications for future research. J. Market., 41–50 (1985)
33. Ricci, M., Parkhurst, G., et al.: Transport policy and social inclusion. Soc. Incl., 1–6 (2016)
34. Turner, J., Grieco, M.: Gender and time poverty: the neglected social policy implications of gendered time, transport, and travel. J. Transp. Geogr., 129–140 (2000)
35. V, M.: Ringing up big business. ORMS Today, 18–24 (1997)
36. Whitt, W.: Dynamic staffing in a telephone call center aiming to immediately answer all calls, 205–212 (1999). Elsevier

A Family of Multi-step Derivative-Free Iterations for Solving Nonlinear Systems

Tugal Zhanlav[1,2], Renchin-Ochir Mijiddorj[3(✉)], and Khuder Otgondorj[2]

[1] Institute of Mathematics and Digital Technology, Mongolian Academy of Sciences, Ulaanbaatar 13330, Mongolia
tzhanlav@mas.ac.mn

[2] School of Applied Sciences, Mongolian University of Science and Technology, Mongolia, Ulaanbaatar 14191, Mongolia

[3] Department of Informatics, Mongolian National University of Education, Peace Avenue, Ulaanbaatar 14191, Mongolia
mijiddorj@msue.edu.mn

Abstract. This paper proposes a family of m-step derivative-free iterative methods for solving nonlinear systems. It is shown that these methods achieve a local convergence order of $2m$ while maintaining low computational cost due to their simple structure. Accelerated variants with memory are also developed and demonstrate improved performance. Numerical experiments confirm the theoretical results and show that the proposed methods are computationally efficient when compared with existing approaches in terms of CPU time.

Keywords: Nonlinear system · Iteration methods · Kurchatov methods · Method with memory

1 Introduction

This paper studies multi-step derivative-free methods for solving nonlinear systems

$$F(x) = 0, \quad x = (x_1, x_2, \ldots, \quad x_n)^T \in R^n. \tag{1}$$

There are numerous derivative free m-step iterations (see [1,6,8,10,13] and references therein). For example, the m-steps scheme proposed in [10] has a simple structure, but its convergence order is low and equal to $(m+1)$. While Cordero et al. considered in [6] SI_m scheme (m-step Symmetric Iterative method) with $2m$ order of convergence. Owing to the complexity of their coefficients, the SWD_m (Steffensen-Wang with Direct inversion) and SWK_m (Steffensen-Wang with Kurchatov acceleration) schemes proposed in [6] are computationally more expensive than the scheme in [10]. Argyros et al. extended SI_m from Euclidean space to the Banach space (a complete normed vector space) [2] and the semilocal analysis of convergence is presented utilizing the majorizing sequence and similar computation. Our aim is to construct families of m-step derivative-free iterative

A. Chinchuluun et al. (Eds.): AAIM 2025, LNCS 16379, pp. 158–175, 2026.
https://doi.org/10.1007/978-981-95-5657-1_13

methods with higher convergence order as scheme SI_m and low computational cost as scheme in [10].

The class of derivative-free methods goes back a long way, e.g., to the quasi-Newton or Broyden-type methods that update an approximation to the Jacobian $\nabla f(x_k)$ by low-rank updates [18–20]. They are particularly well suited for large or sparse systems, and commonly lead to superlinear convergence as they do not recompute an exact Jacobian but rather update a computed one. On the contrary, in this work a new class of purely derivative-free k-step iterations, referred to as symmetric divided difference multi-step formulas with local orders up to $2m$, are proposed which do not require explicit Jacobian update. Therefore, the proposed methods occupy a distinct position between classical quasi-Newton frameworks and high-order Steffensen-type schemes.

2 A Family of Multi-step Derivative-Free Symmetric Iterative Methods

The naming in the derivative free context is used as this meaning in the classical one [6,10,12]. It is in that the algorithms do not need to compute or store the Jacobian matrix (or any high-order derivatives) explicitly. Rather, they use elementary divided difference operators of order one as $[w_k, s_k; F]$ to estimate the Jacobian data based purely on the function values. Even the above listed multi-step schemes are called to be derivative-free, these divided differences act similar to derivative even then retaining its derivative-free approach. All updates are calculated from function values and analytic derivatives not required. This type of difference is particularly interesting in comparison to quasi-Newton or Broyden methods which have an explicit update and store of the approximating Jacobians. By contrast, the current methods avoid updates based on matrix and attain high-order convergence entirely from symmetric divided-difference approximations so that they remain derivative-free but reduce considerably computational cost.

In this section, we introduce a general framework for constructing multi-step derivative-free methods for solving systems of nonlinear equations of the form

$$F(x) = 0, \qquad F : \mathbb{R}^n \to \mathbb{R}^n. \tag{2}$$

The proposed schemes aim to approximate the Jacobian information using divided-difference operators, thereby avoiding explicit derivative evaluation while maintaining high convergence order.

Let $F : \mathbb{R}^n \to \mathbb{R}^n$ be a continuously differentiable function, and let x^* denote a simple zero of F (that is, $F(x^*) = 0$ and $F'(x^*)$ is nonsingular). The following notation and operator conventions are used throughout this paper:

- $x_k \in \mathbb{R}^n$: the k-th iterate;
- $e_k = x_k - x^*$: the error vector at iteration k;
- $F'(x)$: the Jacobian matrix of F evaluated at x;
- I: the identity matrix of appropriate dimension;

- $[w, s; F]$: the first-order divided-difference operator defined by

$$[w, s; F](w - s) = F(w) - F(s), \qquad w, s \in \mathbb{R}^n,\ w \neq s, \tag{3}$$

which approximates $F'(x)$ as $w, s \to x$;
- $\alpha_k, \theta_k \in \mathbb{R}^{n\times n}$: parameter matrices appearing in the construction of the iterative schemes;
- $A_i \in L(\mathbb{R}^n, \mathbb{R}^n)$: linear operators used in intermediate correction steps;
- $\mathbf{x}$, $\mathbf{F}(\mathbf{x})$: vectors (boldface lowercase);
- ψ_j^k: intermediate approximation at step j of iteration k;
- m: number of steps in multi-step methods;
- ϵ: convergence tolerance (typically 10^{-300} in high-precision tests);
- $k_{\max}$: maximum number of iterations.

Consider the iterative process starting from an initial approximation $x_0 \in \mathbb{R}^n$. The next iterate x_{k+1} is generated using a sequence of intermediate approximations $w_k, s_k, \ldots$ and their associated divided-difference operators. Let the first auxiliary point w_k be defined as

$$w_k = x_k - [x_k, x_{k-1}; F]^{-1} F(x_k), \tag{4}$$

which provides a derivative-free analogue of the classical Newton step. Subsequent corrections are introduced through symmetric combinations of divided-difference matrices and weighting parameters α_k and θ_k, yielding a family of multi-step derivative-free symmetric schemes. The general formulation can thus be expressed as

$$x_{k+1} = x_k - M_k F(x_k), \tag{5}$$

where M_k is constructed from suitable combinations of the divided-difference operators and parameter matrices to ensure improved convergence properties.

We consider a family of multi-step derivative-free iterations

$$\begin{aligned}
\psi_1^k &= x_k - [w_k, s_k; F]^{-1} F(x_k), \\
\psi_2^k &= \psi_1^k - \alpha_k [w_k, s_k; F]^{-1} F(\psi_1^k), \\
\psi_3^k &= \psi_2^k - \alpha_k [w_k, s_k; F]^{-1} F(\psi_2^k), \\
&\cdots \\
x_{k+1} = \psi_m^k &= \psi_{m-1}^k - \alpha_k [w_k, s_k; F]^{-1} F(\psi_{m-1}^k),
\end{aligned} \tag{6}$$

where $[w_k, s_k; F]$ denotes the first order symmetric divided difference operator:

$$w_k = x_k + \gamma F(x_k), \quad s_k = x_k - \gamma F(x_k), \quad \gamma \neq 0, \tag{7}$$

i.e.

$$[w_k, s_k; F](w_k - s_k) = F(w_k) - F(s_k). \tag{8}$$

The divided difference operator $[x, y; F]$ associated with a vector-valued function F is defined as an $n \times n$ matrix whose (i, j)-th entry is given by

$$[x, y; F]_{i,j} = \frac{F_i(x_1, \ldots, x_j, y_{j+1}, \ldots, y_n) - F_i(x_1, \ldots, x_{j-1}, y_j, \ldots, y_n)}{x_j - y_j},$$

for $1 \leq i, j \leq n$. To analyze the convergence order of iterations (6), we recall Taylor expansion of function from [10].

Lemma 1. *Let $F : D \subseteq R^n \to R^n$ be $p-$times Fréchet differentiable function in a convex set $D \subseteq R^n$, then for any $x, \hat{h} \in D$ the following expression holds:*

$$F(x+\hat{h}) = F(x) + F'(x)\hat{h} + \frac{1}{2!}F''(x)\hat{h}^2 + \frac{1}{3!}F'''(x)\hat{h}^3 + \cdots + \frac{1}{p!}F^{(p)}(x)\hat{h}^p + R_p,$$

where

$$\|R_p\| \leq \frac{1}{p!} \sup_{0<t<1} \|F^{(p)}(x+t\hat{h})\| \|\hat{h}\|^p \quad \text{and} \quad \hat{h}^p = (\overbrace{\hat{h}, \hat{h}, \cdots, \hat{h}}^{p}) \in R^{n \times p}$$

and $\|\cdot\|$ denotes any norm in R^n, or a corresponding operator norm.

Using expansions of $F(w_k)$ and $F(s_k)$ at point x_k in (8), we obtain

$$[w_k, s_k; F] = F'(x_k) + O\left(F(x_k)^2\right). \tag{9}$$

From (9), it follows that

$$[w_k, s_k; F]^{-1} = F'(x_k)^{-1} + O\left(F(x_k)^2\right). \tag{10}$$

Let $e_k = x_k - \alpha$ denote the error in the $k-th$ approximation of solution α of system (1). Then Taylor expansion of $F(x_k)$ at point α gives [10]

$$F(x_k) = F'(\alpha)\left(e_k + A_2 e_k^2 + A_3 e_k^3\right) + O\left(e_k^4\right). \tag{11}$$

As a consequence of (11), we have

$$F'(x_k) = F'(\alpha)\left(I + 2A_2 e_k + 3A_3 e_k^2\right) + O\left(e_k^3\right), \tag{12}$$

$$F''(x_k) = F''(\alpha)\left(2A_2 + 6A_3 e_k\right) + O\left(e_k^2\right), \tag{13}$$

where $A_i = \frac{1}{i!}F'(\alpha)^{-1}F^{(i)}(\alpha) \subseteq L_i(R^n, R^n)$ and $L_i(R^n, R^n)$ is the set of bounded i-linear functions and $e_k^i = (e_k, e_k, \cdots, e_k)$, $e_k \in R^n$, and $F'(\alpha)^{-1} \subset L(R^n, R^n)$. For iterations (6) holds

Theorem 1. *Let $F : D \subseteq R^n \to R^n$ be a sufficiently Fréchet differentiable function in a convex set $D \subseteq R^n$ containing the simple solution α of $F(x)$. Suppose that $F'(x)$ is continuous and non-singular at α and an initial approximation x_0 is sufficiently close to α. Then the sequence $\{\psi_m^k\}$ obtained by (6) converges to α with order $2m$ if and only if α_k satisfies*

$$\alpha_k = I + 2\theta_k + O\left(F(x_k)^2\right), \tag{14}$$

where

$$\theta_k = \frac{1}{2}F'(x_k)^{-1}F''(x_k)F'(x_k)^{-1}F(x_k). \tag{15}$$

Proof. Let $e\left(\psi_j^k\right) = \psi_j^k - \alpha, \quad j = 1, 2, \cdots, m$. From the first equation in (6), we obtain

$$e\left(\psi_1^k\right) = \psi_1^k - \alpha = e_k - \left[w_k, s_k; F\right]^{-1} F\left(x_k\right).$$

Using (10) and (11) in the last expression, we get

$$e\left(\psi_1^k\right) = \left(I - \left[w_k, s_k; F\right]^{-1} F'(\alpha)\right) e_k + O\left(e_k^2\right). \tag{16}$$

By virtue of (10) and $F'(x_k)^{-1} = F'(\alpha)^{-1} + O(e_k)$, we have

$$\left[w_k, s_k; F\right]^{-1} F'(\alpha) = I + O\left(e_k\right).$$

Then from (16), we have

$$e\left(\psi_1^k\right) = O\left(e_k^2\right)$$

i.e., the assertion is true for $i = 1$. From the second equation in (6), we obtain

$$e\left(\psi_2^k\right) = e\left(\psi_1^k\right) - \alpha_k \left[w_k, s_k; F\right]^{-1} F\left(\psi_1^k\right). \tag{17}$$

Since

$$F\left(\psi_1^k\right) = F'(\alpha) e\left(\psi_1^k\right) + O\left(e_k^4\right)$$

the equation (17) leads to

$$e\left(\psi_2^k\right) = \left[I - \alpha_k \left[w_k, s_k; F\right]^{-1} F'(\alpha)\right] e\left(\psi_1^k\right) + O\left(e_k^4\right).$$

It follows that

$$e\left(\psi_2^k\right) = O\left(e_k^4\right) \tag{18}$$

if and only if

$$I - \alpha_k \left[w_k, s_k; F\right]^{-1} F'(\alpha) = O\left(e_k^2\right)$$

or

$$\alpha_k = F'(\alpha)^{-1} \left[w_k, s_k; F\right] + O\left(e_k^2\right). \tag{19}$$

Using (11), (12), (13), and $F'\left(x_k\right)^{-1} = \left(I - 2A_2 e_k\right) F'(\alpha)^{-1} + O\left(e_k^2\right)$, one can easily shown that for θ_k given by (15) holds

$$\theta_k = A_2 e_k + O\left(e_k^2\right). \tag{20}$$

Using (12) and (20) in (19), we obtain

$$\alpha_k = I + 2\theta_k + O\left(e_k^2\right)$$

i.e., (18) holds if and only if α_k satisfies (14).

By mathematical induction let $e\left(\psi_{m-1}^k\right) = O\left(e_k^{2(m-1)}\right)$. Then from last equation in (6), we obtain

$$e\left(\psi_m^k\right) = e\left(\psi_{m-1}^k\right) - \alpha_k \left[w_k, s_k; F\right]^{-1} F\left(\psi_{m-1}^k\right).$$

Using $F\left(\psi_{m-1}^{k}\right)=F'(\alpha)e\left(\psi_{m-1}^{k}\right)+O\left(e_k^{4(m-1)}\right)$ in the last expression, we get

$$e\left(\psi_{m}^{k}\right)=\left(I-\alpha_k\left[w_k,s_k;F\right]^{-1}F'(\alpha)\right)e\left(\psi_{m-1}^{k}\right)+O\left(e_k^{4(m-1)}\right).$$

By virtue of (14), we have

$$I-\alpha_k\left[w_k,s_k;F\right]^{-1}F'(\alpha)=O\left(e_k^2\right)$$

then we have

$$e_{k+1}=e\left(\psi_{m}^{k}\right)=O\left(e_k^{2\,m}\right).$$

Remark 1. The convergence analysis proposed in this paper is based on the locality of algorithm; i.e., it requires that the initial approximation x_0 be taken sufficiently close to true solution α. Such hypotheses are standard for schemes of higher order of iterative type. It seems to us that the extension of this class from local convergence to semilocal or global convergence is an interesting direction for future work.

Remark 2. Even when the considered iterative schemes are multistep and derivative-free, some of the formulated versions involve divided-difference operators for an estimation of first-order derivatives. These methods are still termed derivative-free in the usual sense, because at no point do we compute (or even store) analytic derivatives or Jacobians; all information comes solely from function evaluations. The divided-difference mechanism provides a numerical approximation that enables high-order convergence while retaining the computational advantages of derivative-free methods.

Note that the iteration in Eq. (6) requires only one matrix $[w_k,s_k;F]$ inversion per iteration step, making it computationally more efficient. The iteration (6) is a derivative-free version of iteration [16,17].

3 Implementation Algorithms for α_k

The expression for θ_k in (15) involves computationally expensive second-order derivatives. We use the notations

$$\begin{aligned}&t_k=F'(x_k)^{-1}F'(y_k),\quad \tilde{t}_k=\left[w_k,s_k;F\right]^{-1}\left[v_k,y_k;F\right],\\&v_k=y_k+cF\left(y_k\right),\quad y_k=\psi_1^k,\quad c\neq 0.\end{aligned}$$

Then it is easy to show that

$$t_k=I-2\theta_k+O\left(F(x_k)^2\right),\qquad \tilde{t}_k=I-2\theta_k+O\left(F(x_k)^2\right).$$

So

$$\alpha_k=I+2\theta_k+O\left(F(x_k)^2\right)=2I-\tilde{t}_k+O\left(F(x_k)^2\right). \tag{21}$$

Another way is to use (12) and (20) in (9) and to obtain

$$[w_k, s_k; F] = F'(\alpha)(I + 2\theta_k) + O\left(F(x_k)^2\right).$$

From this, we find that

$$I + 2\theta_k = F'(\alpha)^{-1}[w_k, s_k; F] + O\left(F(x_k)^2\right). \tag{22}$$

We use expansion of $F'\left(\psi_1^k\right)$ at point α

$$F'\left(\psi_1^k\right) = F'(\alpha) + F''(\alpha)\left(\psi_1^k - \alpha\right) + O\left(\left(\psi_1^k - \alpha\right)^2\right) = F'(\alpha) + O\left(e_k^2\right).$$

Hence,

$$F'(\alpha)^{-1} = F'\left(\psi_1^k\right)^{-1} + O\left(e_k^2\right) = \left[\psi_1^k + cF\left(\psi_1^k\right), \psi_1^k; F\right]^{-1} + O\left(e_k^2\right). \tag{23}$$

Substituting (23) into (22), we obtain

$$\alpha_k = I + 2\theta_k = \left[v_k, \psi_1^k; F\right]^{-1}[w_k, s_k; F] + O\left(e_k^2\right). \tag{24}$$

The iteration (6) with choice α_k given by (24) leads to

$$\begin{aligned}
\psi_1^k &= x_k - [w_k, s_k; F]^{-1} F(x_k), \\
\psi_2^k &= \psi_1^k - \left[v_k, \psi_1^k; F\right]^{-1} F\left(\psi_1^k\right), \quad v_k = \psi_1^k + cF\left(\psi_1^k\right), \ c \neq 0 \\
\psi_3^k &= \psi_2^k - \left[v_k, \psi_1^k; F\right]^{-1} F\left(\psi_2^k\right), \\
&\cdots \\
\psi_m^k &= \psi_{m-1}^k - \left[v_k, \psi_1^k; F\right]^{-1} F\left(\psi_{m-1}^k\right).
\end{aligned} \tag{25}$$

In this case, we obtain m-step derivative-free iterations with scalar coefficients equal to one. Further, in R^n with point-wise multiplication and division of two vectors one can easily shown that [15]

$$\theta_k = \frac{F\left(\psi_1^k\right)}{F(x_k)} + O\left(e_k^2\right). \tag{26}$$

Substituting the approximation (26) into (14), we obtain the simplified coefficient

$$\alpha_k = \mathbf{1} + 2 \cdot \frac{F\left(\psi_1^k\right)}{F(x_k)} + O\left(e_k^2\right), \qquad \mathbf{1} = (1, 1, \ldots, 1). \tag{27}$$

In this case iterations (6) can be considered as iteration with vector coefficients (27). Thus, we prove three different implementation algorithms (21), (24) and (27) for α_k. We denote by SI_m the family of iterations (6) with w_k and s_k as defined in (7). In fact, we develop three different m-step derivative-free iterations. The specificity of iterations (25) is that it requires two inverses of matrices $[w_k, s_k; F]$ and $\left[v_k, \psi_1^k; F\right]$ and does not require time consuming coefficients calculations. On the contrary, the relations (6) with choices (21) and (27) require only one inverse of matrix $[w_k, s_k; F]$ and require time consuming coefficients calculations.

4 Derivative Free Symmetric Iteration with Memory

Recently, Ezquerro et al. [7] proposed uniparametric family of iterations with memory and with quadratic convergence

$$x_{k+1} = x_k - [w_k, s_k; F]^{-1} F(x_k), \qquad k = 0, 1, \ldots, \tag{28}$$

where

$$w_k = x_k + \beta (x_k - x_{k-1}), \quad s_k = x_k - \beta (x_k - x_{k-1}), \quad \beta \in (0, 1). \tag{29}$$

The family (28) contains Kurchatov's method and Newton's method, since (28) is reduced to Kurchatov's method if $\beta = 1$ and to Newton's method if $\beta = 0$ and F is differentiable. Regarding Kurchatov's method, many researchers have devoted and investigated this problem. For example Argyros et al. [1] and Kumar et al. [9] investigated the local and semilocal convergence of two-step Kurchatov-type methods under conditions that the first-order divided difference satisfy the generalized Lipschitz conditions. Regions of convergence and uniqueness of the solutions are established. But in the above mentioned works do not establish the convergence order. They concluded that the method converges more rapidly than the one-step Kurchatov method and two-step secant method. Cordero et al. [4] studied a family of one-step Kurchatov and its extension for multidimensional case. Chicharro et al. [3] and Wang et al. [12] considered higher-order derivative-free variants of Kurchatov-type methods with accelerating parameters. Here, we consider m-step derivative-free iterations (6) with w_k, s_k given by formula (29). As in the preceding case it is easy to check that the formulae (9) and (10) hold true. Therefore, Theorem 1 also holds in this case. We denote iterations (6) with w_k, s_k given by formula (29) by SIM_m. Moreover, the implementation algorithms (21), (24), and (27) also perform well for SIM_m.

Remark 3. The proposed SIM_m can be regarded as multi-step extension of Kurchatov's method or method (28).

Note that in [7] proven quadratic convergence of family (28) for $\beta \in (0, 1)$. But if we consider (28) as independent iterative methods or continuous analogy of Newton's method [14] then better choice of parameter is $\beta \in (0, 2)$. The pseudocodes of the proposed SI and SIM algorithms are included in the Appendix for reference.

5 Numerical Experiments

This section presents high-precision numerical tests illustrating the performance of the proposed multi-step methods SI_m and SIM_m, both in their basic and memory-enhanced forms. All computations were carried out in *Wolfram Mathematica 14* using 10 000-digit arbitrary-precision arithmetic on a 1.30 GHz Intel(R) Core(TM) HP laptop. Unless otherwise stated, default parameter values were $c = \beta = \gamma = \delta = 1/10$.

Table 1. Numerical results for Example 1: comparison of iteration count, CPU time, and ACOC for various methods.

Methods	$\|x_{k+1} - x_k\|$	$\|F(x_{k+1})\|$	Iter.	ACOC	Time (s)
SW_1	1.15550×10^{-438}	1.27019×10^{-878}	9	2.000	54
SWD_1	1.80864×10^{-392}	1.76568×10^{-949}	7	2.414	86
SWK_1	8.41635×10^{-643}	$2.17986 \times 10^{-1757}$	7	2.732	95
SI_1 (6)	1.07996×10^{-454}	9.74248×10^{-911}	9	2.000	55
SIM_1 (28), (29)	2.62971×10^{-444}	1.16671×10^{-889}	9	2.000	57
SW_2	5.16946×10^{-508}	$3.18774 \times 10^{-2036}$	5	4.000	65
SWD_2	4.37888×10^{-790}	$3.00129 \times 10^{-3520}$	5	4.449	85
SWK_2	8.08523×10^{-968}	$6.33371 \times 10^{-4677}$	5	4.831	104
SI_2 (6), (21)	3.76595×10^{-365}	$2.95463 \times 10^{-1464}$	5	4.000	61
SIM_2 (6), (21)	9.55003×10^{-365}	$3.01424 \times 10^{-1463}$	5	4.000	60
SI_2 (25), (24)	3.97489×10^{-450}	$8.59112 \times 10^{-1805}$	5	4.000	50
SIM_2 (25), (24)	1.89937×10^{-450}	$4.47908 \times 10^{-1806}$	5	4.000	61
SI_2 (6), (27)	$8.10220 \times 10^{-1138}$	$2.86896 \times 10^{-4553}$	6	4.000	37
SIM_2 (6), (27)	1.89861×10^{-576}	$1.91554 \times 10^{-1869}$	6	3.240	39
SW_3	1.33840×10^{-407}	$2.26096 \times 10^{-2452}$	4	6.000	55
SWD_3	1.52722×10^{-548}	$9.10959 \times 10^{-3553}$	4	6.461	83
SWK_3	2.18337×10^{-610}	$7.30139 \times 10^{-4197}$	4	6.862	75
SI_3 (6), (21)	$4.45222 \times 10^{-1738}$	$4.84365 \times 10^{-10017}$	5	6.000	49
SIM_3^* (6), (21)	$4.52427 \times 10^{-1708}$	$2.48100 \times 10^{-10017}$	5	6.000	57
SI_3 (25), (24)	1.62263×10^{-367}	$4.86044 \times 10^{-2212}$	4	6.000	49
SIM_3 (25), (24)	8.71957×10^{-368}	$1.17037 \times 10^{-2213}$	4	6.000	48
SI_3 (6), (27)	$1.07073 \times 10^{-1290}$	$8.03171 \times 10^{-7747}$	5	6.000	27
SIM_3 (6), (27)	2.75570×10^{-789}	$1.22652 \times 10^{-3815}$	5	4.826	31

SIM_3^*, where $\tilde{t}_k = [w_k, s_k; F]^{-1} [v_k, y_k; F], \quad v_k = y_k - [w_k, s_k; F]^{-1} F(y_k)$

Iterations were terminated once

$$\|x_{k+1} - x_k\|_2 + \|F(x_{k+1})\|_2 < 10^{-300}.$$

The empirical convergence order was estimated using the Approximate Computational Order of Convergence (ACOC) introduced by Cordero and Torregrosa [5]:

$$p \approx \text{ACOC} = \frac{\ln(\|x_{k+1} - x_k\|_2 / \|x_k - x_{k-1}\|_2)}{\ln(\|x_k - x_{k-1}\|_2 / \|x_{k-1} - x_{k-2}\|_2)}.$$

Example 1. Following Narang et al. [10], we solve

$$\begin{cases} x_i \sin(x_i + 1) - 1 = 0, & 1 \le i \le 49, \\ x_{50} \sin(x_1) - 1 = 0, \end{cases}$$

starting from the uniform guess $x_0 = \{13/10, 13/10, ..., 13/10\}^T$. The reference root is
$\alpha = \{1.1141571408719300..., 1.1141571408719300..., ..., 1.1141571408719300...\}^T$.

Notation such as SI_2 (6), (21) or SIM_2 (6), (21), used in Tables 1, 2 and 3, denotes the application of iteration scheme (6) with α_k computed via (21). The scheme differ in their divided difference definitions: SI uses equation (7), while SIM employs equation (29). The numerical results obtained by the methods SW_m, SWD_m, and SWK_m for solving systems are presented in Tables 1, 2 and 3, following the approach proposed in [6]. In addition, for the methods with memory SWD_m and SWK_m, we use as initial approximation x_{-1}, vector $[1.4, ..., 1.4]^T$.

Tables 1 and 2 report the norm of the last update, residual norm, iteration count and CPU time. The numerical results can be seen in Table 1. The proposed SI_2 method achieves the same tolerance (4th-order convergence) as SW_2 in 5 iterations, but with 6.2% faster CPU time (61 s vs 65 s). SIM_3 (25), (24) with memory reduces CPU time by 36% compared to SWK_3 (48 s vs 75 s) while maintaining 6th-order convergence. As shown in Fig. 1, the proposed SI and SIM methods exhibit stable and efficient convergence.

Example 2. Next, we consider a system of equations for $n = 250$:

$$\begin{cases} x_i^2 x_{i+1} - 1 = 0, & i = 1, 2, \ldots, 249, \\ x_{250}^2 x_1 - 1 = 0. \end{cases}$$

The solution of the given system of equations is $\alpha = \{1, 1, \ldots, 1\}^T$ and the initial guess is taken as $x_0 = \{2, 2, \ldots, 2\}^T$.

The numerical results can be seen in Table 2. SI_2 (6), (27) vs SWD_2: Despite having identical ACOC (4.0), SI_2 with vector coefficients (27) is 52.9% faster than SWD_2 (85 s vs 175 s). SIM_3 (6), (21) vs SWK_3: SIM_3 requires 23% less CPU time than SWK_3 (139 s vs 181 s) for 6th-order convergence.

Example 3. Consider the solution of a nonlinear partial differential equation using finite difference discretization, which leads to a system of nonlinear algebraic equations. We focus on solving Fisher's equation (see [11]), reaction–diffusion equation with homogeneous Neumann boundary conditions given by:

$$\begin{cases} u_t = Du_{xx} + u(1-u) \\ u(x,0) = 1.5 + 0.5\cos(\pi x), & 0 \le x \le 1 \\ u_x(0,t) = 0, & \forall t \ge 0 \\ u_x(1,t) = 0, & \forall t \ge 0. \end{cases}$$

Here, $D = 1$ is the diffusion coefficient. Let M and N denote the number grid points in the spatial x and time t directions ($0 \le t \le 3$), respectively, with step sizes $h = \triangle x$ and $k = \triangle t$. The exact solution is denoted by $u = u(x,t)$, and its numerical approximation at grid point (x_i, t_j) is $w_{i,j} \simeq u(x_i, t_j)$.

We discretize the PDE using:
Backward difference for the time derivative:

$$u_t(x_i,t_j) \simeq \frac{(w_{i,j}-w_{i,j-1})}{k}.$$

Central differences for the spatial derivatives:

$$u_x(x_i,t_j) \simeq \frac{(w_{i+1,j}-w_{i-1,j})}{2h}, \quad u_{xx}(x_i,t_j) \simeq \frac{(w_{i+1,j}-2w_{i,j}+w_{i-1,j})}{h^2}.$$

For $M = 21$ and $N = 21$, (resulting in $m = M - 1 = 20$, $n = N - 1 = 20$), the discretization yields a system of 400 nonlinear equations. The initial guess for the solution is chosen as:

$$x_0 = \text{Table}[3 + \frac{i}{m \cdot n}, \{i, 1, m \cdot n\}].$$

This system is then solved using various numerical methods.

Table 2. Numerical results for Example 2: comparison of iteration count, CPU time, and ACOC for various methods.

Methods	$\|x_{k+1}-x_k\|$	$\|F(x_{k+1})\|$	Iter.	ACOC	Time (s)
SW_1	2.84017×10^{-567}	$1.98968 \times 10^{-1134}$	13	2.000	199
SWD_1	2.12916×10^{-448}	$1.01358 \times 10^{-1082}$	9	2.414	192
SWK_1	5.23099×10^{-320}	6.60848×10^{-875}	8	2.732	220
SI_1 (6)	8.84724×10^{-484}	1.48514×10^{-967}	12	2.000	129
SIM_1 (28), (29)	4.11634×10^{-503}	$3.22562 \times 10^{-1003}$	12	2.000	127
SW_2	3.79774×10^{-484}	$3.46852 \times 10^{-1937}$	7	4.000	201
SWD_2	3.36565×10^{-559}	$5.74480 \times 10^{-2489}$	6	4.449	175
SWK_2	8.64354×10^{-629}	$1.86196 \times 10^{-3037}$	6	4.827	196
SI_2 (6), (21)	4.91363×10^{-609}	$2.07931 \times 10^{-2436}$	7	4.000	167
SIM_2 (6), (21)	4.25772×10^{-646}	$9.99319 \times 10^{-2585}$	7	4.000	155
SI_2 (25), (24)	3.15525×10^{-819}	$9.77891 \times 10^{-3278}$	7	4.000	129
SIM_2 (25), (24)	1.89800×10^{-855}	$1.28039 \times 10^{-3422}$	7	4.000	153
SI_2 (6), (27)	6.27868×10^{-610}	$5.46879 \times 10^{-2440}$	7	4.000	85
SIM_2 (6), (27)	1.77532×10^{-353}	$2.78874 \times 10^{-1146}$	7	3.241	77
SW_3	1.31083×10^{-792}	$1.01171 \times 10^{-4756}$	6	6.000	170
SWD_3	1.19386×10^{-509}	$8.09848 \times 10^{-3296}$	5	6.464	181
SWK_3	6.71249×10^{-486}	$1.56884 \times 10^{-3342}$	5	6.875	181
SI_3 (6), (21)	1.38121×10^{-985}	$5.64670 \times 10^{-5914}$	6	6.000	128
SIM_3^* (6), (21)	$1.50917 \times 10^{-1045}$	$7.17356 \times 10^{-6274}$	6	6.000	139
SI_3 (25), (24)	$3.44443 \times 10^{-1318}$	$1.51582 \times 10^{-7910}$	6	6.000	124
SIM_3 (25), (24)	$1.39963 \times 10^{-1379}$	$6.82373 \times 10^{-8279}$	6	6.000	123
SI_3 (6), (27)	5.66248×10^{-987}	$2.61541 \times 10^{-5922}$	6	6.000	58
SIM_3 (6), (27)	8.32594×10^{-681}	$2.97879 \times 10^{-3290}$	6	4.824	57

SIM_3^*, where $\tilde{t}_k = [w_k, s_k; F]^{-1}[v_k, y_k; F]$, $v_k = y_k - [w_k, s_k; F]^{-1} F(y_k)$.

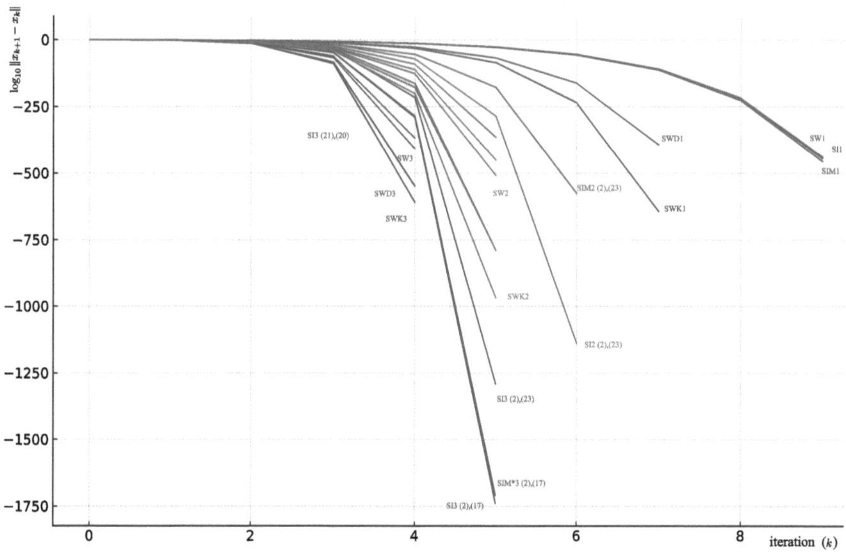

Fig. 1. Convergence history for Example 1 comparing the proposed SI and SIM methods with existing ones.

The numerical results can be seen in Table 3. For Example 3, CPU-time results are not included because the computations were performed using high-precision arithmetic with 6 000 to 20 000 digits. The primary objective of this experiment was to illustrate the order of convergence, rather than computational speed.

6 Discussion and Robustness Analysis

The analysis of convergence provided above, which is local in nature, requires that an initial approximation x_0 be placed relatively close to the exact root α. In practice, in the case of derivative-free multi-step methods, there is no clear limit on their performance because it can differ greatly depending on the initial guess and conditioning of the system.

Sensitivity to Initial Guesses

We tested for robustness concerning the starting point by performing additional tests in which we randomly perturbed the start vectors around the reference solution. The results indicate that while all proposed methods converge quadratically or super-quadratically when x_0 lies inside the local basin of attraction, the width of this basin shrinks as the convergence order $2m$ increases. Consequently, higher-order variants such as SWK_3 or SIM_3 may diverge or oscillate if the initial guess is far from α. Such a behavior is typical of high-order schemes and highlights the significance for scaling or damping approaches in use.

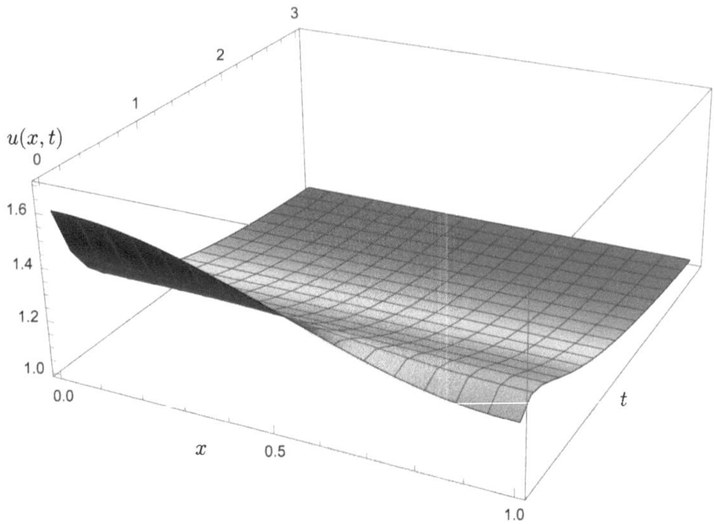

Fig. 2. Numerical solution of Fisher's equation obtained via SI_3 (6), (27).

Ill-Conditioned and Noisy Systems

Then in the presence of a nearly-singular Jacobian $F'(x)$ (or an ill-conditioned system) if small variations of $F(x)$ can be significantly magnified, which may result in loss of accuracy or even numerical instability.

To mitigate these issues, practical implementations may combine the present iterations with a line-search or trust-region safeguard ensuring monotonic residual decrease:

$$F(x_k + s_k) \leq F(x_k) - c\,\|s_k\|^2, \quad c \in (0, 1).$$

Alternatively, a damping or Levenberg–Marquardt-type correction

$$\big([w_k, s_k; F]^T [w_k, s_k; F] + \lambda I\big) s_k = -[w_k, s_k; F]^T F(x_k)$$

can improve robustness at the cost of one extra matrix solve per iteration.

Limitations and Further Directions

Despite the fact that the proposed schemes are very accurate on smooth well-conditioned problems, their local nature makes them not suitable to be used in non-ideal or discontinuous environments. Adaptive pre-conditioning, auto-stepping or hybrid switching between derivative-free vs quasi-Newton updates as promising directions for improving global robustness. A systematic semilocal analysis extending Theorem 1 to the ill-conditioned case remains an open topic for future research.

In order to deal with the scalability issues in very large-scale settings (e.g., million-variable systems), our proposed procedures can be generalized by considering that an adaptive precision. While ultra-high precision (10 000 digits) was

Table 3. Numerical results for Example 3: comparison of iteration count and ACOC for various methods.

Methods	$\|x_{k+1} - x_k\|$	$\|F(x_{k+1})\|$	Iter.	ACOC
SW_1	1.12945×10^{-436}	6.94532×10^{-876}	12	2.000
SWD_1	2.96806×10^{-539}	$9.27199 \times 10^{-1305}$	9	2.414
SWK_1	2.77266×10^{-472}	$3.15494 \times 10^{-1420}$	8	3.000
SI_1 (6)	2.69639×10^{-436}	3.96638×10^{-875}	12	2.000
SIM_1 (28), (29)	2.69639×10^{-436}	3.96638×10^{-875}	12	2.000
SW_2	3.48448×10^{-743}	$5.88432 \times 10^{-2977}$	7	4.000
SWD_2	4.39931×10^{-653}	$3.83971 \times 10^{-2910}$	6	4.449
SWK_2	$7.11457 \times 10^{-1231}$	$1.53342 \times 10^{-7390}$	6	6.000
SI_2 (6), (21)	2.45224×10^{-537}	$7.26379 \times 10^{-2153}$	7	4.000
SIM_2 (6), (21)	4.01654×10^{-580}	$4.18057 \times 10^{-2324}$	7	4.000
SI_2 (25), (24)	1.06847×10^{-873}	$5.22324 \times 10^{-3499}$	7	4.000
SIM_2 (25), (24)	1.06847×10^{-873}	$5.22324 \times 10^{-3499}$	7	4.000
SI_2 (6), (27)	4.30680×10^{-533}	$4.16359 \times 10^{-1602}$	8	3.005
SIM_2 (6), (27)	4.30680×10^{-533}	$4.16359 \times 10^{-1602}$	8	3.005
SW_3	$2.69490 \times 10^{-1222}$	$2.24417 \times 10^{-7339}$	6	6.000
SWD_3	1.45258×10^{-603}	$3.56721 \times 10^{-3907}$	5	6.464
SWK_3	6.07566×10^{-919}	$1.33195 \times 10^{-8277}$	5	9.000
SI_3 (6), (21)	6.47098×10^{-890}	$6.51580 \times 10^{-5344}$	6	6.000
SIM_3^* (6), (21)	6.37953×10^{-968}	$3.98539 \times 10^{-5812}$	6	6.001
SI_3 (25), (24)	$1.58882 \times 10^{-1418}$	$9.48133 \times 10^{-8517}$	6	6.000
SIM_3 (25), (24)	$1.58882 \times 10^{-1418}$	$9.48133 \times 10^{-8517}$	6	6.000
SI_3 (6), (27)	4.74651×10^{-798}	$1.75695 \times 10^{-3184}$	7	4.079
SIM_3 (6), (27)	4.74651×10^{-798}	$1.75695 \times 10^{-3184}$	7	4.079

SIM_3^*, where $\tilde{t}_k = [w_k, s_k; F]^{-1} [v_k, y_k; F]$, $v_k = y_k - [w_k, s_k; F]^{-1} F(y_k)$

used to verify theoretical convergence, practical implementations would require only moderate precision (50–100 digits) for most computations, reserving higher accuracy for numerically sensitive steps such as matrix inversions. This approach preserves high-order convergence while ensuring computational efficiency and feasible memory use. Future work will explore hybrid and preconditioned extensions to improve performance for sparse or structured large-scale problems.

Conclusion

We introduce families of derivative-free m-step iterations with higher order and low computational complexity. We show that accelerated memories are not only feasible, but their use is also highly effective. The family of the methods without memory, being suggested, contains Kurchatov's one as the particular case.

Algorithm 1. Basic SI_m derivative-free method

Require: Initial guess x_0, function $F : \mathbb{R}^n \to \mathbb{R}^n$, parameters γ, c, steps m, tolerance ϵ

Ensure: Solution x^* such that $\|x_{k+1} - x_k\| + \|F(x^*)\| < \epsilon$

$k \leftarrow 0,\ x_k \leftarrow x_0$
while $k < k_{\max}$ **do**
  Compute $F_k \leftarrow F(x_k)$
  Compute $w_k \leftarrow x_k + \gamma F_k,\ s_k \leftarrow x_k - \gamma F_k$
  Compute $F_w \leftarrow F(w_k),\ F_s \leftarrow F(s_k)$
  Construct divided difference $[w_k, s_k; F]$ using Definition (8)
  Compute $\psi_1 \leftarrow x_k - [w_k, s_k; F]^{-1} F_k$
  $j \leftarrow 1$
  while $j < m$ **do**
    $\alpha_k \leftarrow$ **Algorithm3**$(x_k, \psi_1, F, \text{methodType})$
    Compute $F_{\psi_j} \leftarrow F(\psi_j)$
    $\psi_{j+1} \leftarrow \psi_j - \alpha_k [w_k, s_k; F]^{-1} F_{\psi_j}$
    $j \leftarrow j + 1$
  end while
  $x_{k+1} \leftarrow \psi_m$
  if $\|x_{k+1} - x_k\| + \|F(x_{k+1})\| < \epsilon$ **then break**
  end if
  $x_k \leftarrow x_{k+1},\ k \leftarrow k + 1$
end while
return x_k

Declarations

Funding. This work was partially supported by the Natural Science Foundation of Inner Mongolia (2023MS06003, 2023KFGJ03).

Data availability. This manuscript has no associated data.

Ethical Conduct. Not applicable.

Conflict of interest. The authors declare no conflict of interest.

Algorithm 2. Memory-enhanced SIM_m method

Require: Initial guesses x_0, x_{-1}, function $F : \mathbb{R}^n \to \mathbb{R}^n$, parameters β, c, steps m, tolerance ϵ
Ensure: Solution x^* such that $\|x_{k+1} - x_k\| + \|F(x^*)\| < \epsilon$

$k \leftarrow 0,\ x_k \leftarrow x_0,\ x_{k-1} \leftarrow x_{-1}$
while $k < k_{\max}$ **do**
  Compute $F_k \leftarrow F(x_k)$
  Compute $w_k \leftarrow x_k + \beta(x_k - x_{k-1}),\ s_k \leftarrow x_k - \beta(x_k - x_{k-1})$
  Compute $F_w \leftarrow F(w_k),\ F_s \leftarrow F(s_k)$
  Construct divided difference $[w_k, s_k; F]$ using Definition (8)
  Compute $\psi_1 \leftarrow x_k - [w_k, s_k; F]^{-1} F_k$
  $j \leftarrow 1$
  while $j < m$ **do**
    $\alpha_k \leftarrow$ **Algorithm3**$(x_k, \psi_1, F, \text{methodType})$
    Compute $F_{\psi_j} \leftarrow F(\psi_j)$
    $\psi_{j+1} \leftarrow \psi_j - \alpha_k [w_k, s_k; F]^{-1} F_{\psi_j}$
    $j \leftarrow j + 1$
  end while
  $x_{k+1} \leftarrow \psi_m$
  if $\|x_{k+1} - x_k\| + \|F(x_{k+1})\| < \epsilon$ **then break**
  end if
  $x_{k-1} \leftarrow x_k,\ x_k \leftarrow x_{k+1},\ k \leftarrow k + 1$
end while
return x_k

Algorithm 3. Computation of parameter α_k

Require: Current iterates x_k, ψ_1, function F
Ensure: Parameter matrix α_k

Method 1: Using equation (21)
Compute $y_k \leftarrow \psi_1$
Compute $v_k \leftarrow y_k + cF(y_k)$
Compute $[v_k, y_k; F]$ using Definition (8)
Compute $\tilde{t}_k \leftarrow [w_k, s_k; F]^{-1} [v_k, y_k; F]$
$\alpha_k \leftarrow 2I - \tilde{t}_k$
Method 2: Using equation (24)
Compute $y_k \leftarrow \psi_1$
Compute $v_k \leftarrow y_k + cF(y_k)$
Compute $[v_k, y_k; F]$ using Definition (8)
$\alpha_k \leftarrow [v_k, y_k; F]^{-1} [w_k, s_k; F]$
Method 3: Using equation (27)
Compute $F_{\psi_1} \leftarrow F(\psi_1),\ F_{x_k} \leftarrow F(x_k)$
Compute vector division: $\theta_k \leftarrow F_{\psi_1} / F_{x_k}$ (element-wise)
$\alpha_k \leftarrow \mathbf{1} + 2\theta_k$ where $\mathbf{1} = (1, 1, \ldots, 1)^T$
return α_k

References

1. Argyros, I.K., Shakhno, S., Regmi, S., Yarmola, H.: On the convergence of two-step Kurchatov-type methods under generalized continuity conditions for solving nonlinear equations. Symmetry **14**(12), 2548 (2022). https://doi.org/10.3390/sym14122548
2. Argyros, I.K., Shakhno, S., Regmi, S., Yarmola, H., Argyros, M.I.: Symmetric-type multi-step difference methods for solving nonlinear equations. Symmetry **16**(3), 330 (2024). https://doi.org/10.3390/sym16030330
3. Chicharro, F.I., Cordero, A., Garrido, N., Torregrosa, J.R.: On the improvement of the order of convergence of iterative methods for solving nonlinear systems by means of memory. Appl. Math. Lett. **104**, 106277 (2020). https://doi.org/10.1016/j.aml.2020.106277
4. Cordero, A., Soleymani, F., Torregrosa, J.R., Haghani, F.K.: A family of Kurchatov-type methods and its stability. Appl. Math. Comput. **294**, 264–279 (2017). https://doi.org/10.1016/j.amc.2016.09.021
5. Cordero, A., Torregrosa, J.R.: Variants of Newton's method using fifth-order quadrature formulas. Appl. Math. Comput. **190**, 686–698 (2007). http://dx.doi.org/10.1016/j.amc.2007.01.062
6. Cordero, A., Villalba, E.G., Torregrosa, J.R., Triguero-Navarro, P.: Introducing memory to a family of multi-step multidimensional iterative methods with weight function. Expositiones Mathematicae **41**, 398–417 (2023). https://doi.org/10.1016/j.exmath.2023.04.004
7. Ezquerro, J.A., Hernández-Verón, M.A., Magreñán, Á.A., Moysi, A.: A procedure to obtain quadratic convergence from the secant method. J. Comput. Appl. Math. **448**, 115912 (2024). https://doi.org/10.1016/j.cam.2024.115912
8. Kansal, M. Rani, L.: An adaptive Steffensen-like families for solving nonlinear systems using frozen divided differences. J. Math. Chem. **62**, 109–144 (2024). https://doi.org/10.1007/s10910-023-01524-1
9. Kumar, H., Parida, P.K.: On semilocal convergence of two step Kurchatov method. Int. J. Comput. Methods **96** (8), 1548–1566 (2019). https://doi.org/10.1080/00207160.2018.1428741
10. Narang, M., Bhatia, S., Alshomrani, A.S., Kanwar, V.: General efficient class of Steffensen type methods with memory for solving systems of nonlinear equations. J. Comput. Appl. Math. **352**, 23–39 (2019). https://doi.org/10.1016/j.cam.2018.10.048
11. Sauer, T.: Numerical Analysis, 2nd edn. Pearson, USA (2012)
12. Wang, X., Jin, Y., Zhao, Y.: Derivative-free iterative methods with some Kurchatov-type accelerating parameters for solving nonlinear systems. Symmetry **13**(6), 943 (2021). https://doi.org/10.3390/sym13060943
13. Zhanlav, T., Mijiddorj, R.: On the connection between high-order iterations with and without derivatives for solving systems of nonlinear equations. J. Math. Sci. (2025). https://doi.org/10.1007/s10958-025-07862-6
14. Zhanlav, T., Chuluunbaatar, O.: New developments of Newton-type iterations for solving nonlinear problems. Springer Nature, Switzerland (2024). https://link.springer.com/book/10.1007/978-3-031-63361-4
15. Zhanlav, T., Otgondorj, Kh.: Development and adaptation of higher order iterative methods in R^n with specific rules. Discrete Continuous Model. Appl. Computat. Sci. **32**(4), 425–444 (2024). https://journals.rudn.ru/miph/article/view/43671

16. Zhanlav, T., Otgondorj, Kh.: Higher order Jarratt-like iterations for solving systems of nonlinear equations. Appl. Math. Comput. **395**, 125849 (2021). https://doi.org/10.1016/j.amc.2020.125849
17. Zhanlav, T., Chun, Ch., Otgondorj, Kh.: Construction and dynamics of efficient high-order methods for nonlinear systems. Int. J. Comput. Meth. **19**, 2250020-1–31 (2022). https://doi.org/10.1142/S0219876222500207
18. Broyden, C. G.: A class of methods for solving nonlinear simultaneous equations. Math. Comp. **19**(92), 577–593 (1965). https://www.jstor.org/stable/2003941
19. Dennis, J. E., Schnabel, R. B.: Numerical Methods for Cnconstrained Optimization and Nonlinear Equations. Prentice-Hall (1983)
20. Kelley, C. T.: Iterative Methods for Linear and Nonlinear Equations. SIAM (1995)

A Novel Nonstandard Finite Difference Scheme for the SIS Epidemiological Model

Tumurkhuyag Badarch[1,2] and Batgerel Balt[1(✉)]

[1] Institute of Mathematics and Digital Technology, Mongolian Academy of Sciences, Ulaanbaatar 13330, Mongolia
{tumurkhuyag.b,batgerel}@mas.ac.mn

[2] School of Information Technology and Electronics, National University of Mongolia, Ulaanbaatar 14200, Mongolia

Abstract. The SIS model is one of the fundamental frameworks used to describe infectious diseases in which recovered individuals do not acquire permanent immunity. However, traditional numerical schemes such as the Euler and Runge–Kutta methods often fail to preserve key qualitative properties of the model—namely, the positivity of solutions, conservation of total population, and dynamic consistency—especially when large time steps are used. In this study, we propose a novel Nonstandard Finite Difference (NSFD) scheme for the SIS model. The scheme guarantees the positivity of solutions, conservation of total population, and elementary stability, while exhibiting weak dependence on both the step size and parameter variations. Theoretical analysis and numerical experiments demonstrate that the proposed scheme provides more stable, robust, and dynamically consistent solutions than the standard Euler, second-order Runge–Kutta (RK2), and Mickens-type NSFD methods. The results indicate that the proposed NSFD approach offers an accurate and reliable numerical tool for simulating SIS and other structurally similar epidemic models.

Keywords: Nonstandard finite difference scheme · SIS epidemic model · Positivity · Population conservation · Elementary stability · Numerical simulation

1 Introduction

Mathematical models of infectious disease transmission play a crucial role in understanding epidemic mechanisms, control strategies, and prevention measures. Among these, the SIS (Susceptible–Infected–Susceptible) model is a fundamental framework used to describe diseases for which recovered individuals do not acquire permanent immunity and thus become susceptible again. Such models are particularly relevant for infections such as gonorrhea or the common cold. To simulate these models numerically, classical finite difference schemes such as the forward Euler method are often employed. However, these standard approaches tend to lose essential qualitative properties of the continuous

A. Chinchuluun et al. (Eds.): AAIM 2025, LNCS 16379, pp. 176–188, 2026.
https://doi.org/10.1007/978-981-95-5657-1_14

system—such as the positivity of solutions, conservation of total population, and dynamic consistency—when the time step becomes relatively large. In addition, conventional schemes are often highly sensitive to parameter variations, which can distort the system's stability and long-term behavior. Hence, the loss of dynamic consistency arises not only from large step sizes but also from parameter sensitivity inherent in the numerical scheme. To address these limitations, *Nonstandard Finite Difference* schemes, initially proposed by Mickens [1,2], have been widely studied. NSFD methods provide a flexible framework for constructing discretizations that preserve the essential qualitative features and dynamic consistency of the original continuous model. These methods have been successfully applied to various epidemic models, maintaining non-negativity, boundedness, and correct steady-state behavior [3–5].

In parallel, stochastic differential equation (SDE)-based approaches and network-structured epidemic models have recently gained significant attention [6–10]. These models incorporate randomness, social contact structures, and mobility patterns, offering more realistic representations of disease transmission. Such high-dimensional and structured systems require numerical schemes that ensure positivity, stability, and reliable long-term behavior. Despite these advances, there remains a strong need for numerical schemes that are both accurate and stable for the SIS model, while being weakly dependent on the step size and parameter variations.

In this study, we develop a novel nonstandard finite difference scheme for the SIS model. The proposed method guarantees positivity, conservation of total population, and elementary stability, while maintaining dynamic consistency with weak sensitivity to both the time-step size and model parameters. Theoretical and numerical analyses demonstrate that the proposed NSFD scheme produces more stable, accurate, and dynamically consistent solutions than the standard Euler, second-order Runge–Kutta, and Mickens-type NSFD methods. These results highlight the reliability and effectiveness of the proposed approach for the numerical simulation of SIS-type and structurally similar epidemic models.

2 The SIS Model

We consider the classical SIS model, which describes the transmission dynamics of infectious diseases that do not confer long-lasting immunity upon recovery. In this model, the total population is divided into two compartments: the susceptible class $S(t)$ and the infectious class $I(t)$, for time $t \geq 0$. Populations who recover from infection return directly to the susceptible class and may be reinfected (Fig. 1).

The model is governed by the following system of ordinary differential equations:

$$\begin{aligned} \frac{dS}{dt} &= \Lambda - \beta SI - \mu S + \xi I, \\ \frac{dI}{dt} &= \beta SI - (\mu + \xi)I, \end{aligned} \tag{1}$$

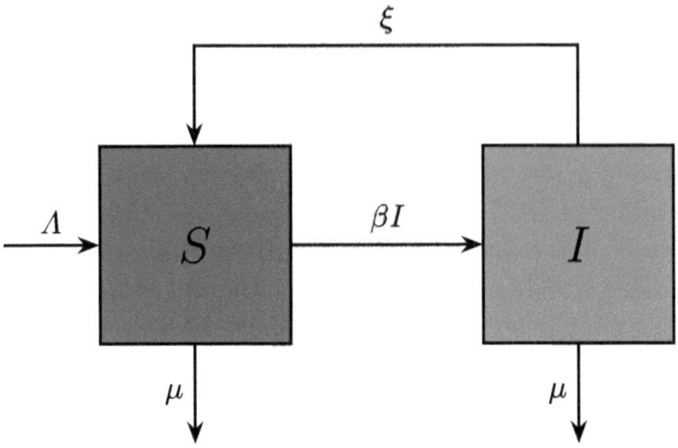

Fig. 1. Flow diagram of the SIS model.

where the parameters are defined as follows:

- Λ is the recruitment (birth) rate,
- β is the transmission coefficient,
- μ is the natural death rate,
- ξ is recovery rate (i.e., recovery followed by return to susceptibility),
- $S(t)$ and $I(t)$ represent the number of susceptible and infectious populations at time t, respectively.

The basic reproduction number $\mathcal{R}_0$ for system (1) is given by:

$$\mathcal{R}_0 = \frac{\Lambda}{\mu} \cdot \frac{\beta}{\mu + \xi}. \tag{2}$$

The system admits two equilibria depending on the value of $\mathcal{R}_0$:

- If $\mathcal{R}_0 \leq 1$, the system has a unique disease-free equilibrium:

$$E_0^* = \left(\frac{\Lambda}{\mu}, 0\right), \tag{3}$$

  which is globally asymptotically stable. In this state, the disease dies out, and the entire population remains susceptible.
- If $\mathcal{R}_0 > 1$, the disease-free equilibrium becomes unstable, and the system admits a unique endemic equilibrium:

$$E_1^* = \left(\frac{\mu + \xi}{\beta}, \frac{\mu + \xi}{\beta} \cdot (\mathcal{R}_0 - 1)\right), \tag{4}$$

  which is locally and globally asymptotically stable. In this case, the infection persists in the population at a steady state.

3 Nonstandard Finite Difference Scheme for the SIS Model

In this section, we develop the NSFD scheme for the SIS model defined in system (1). The primary goal of the NSFD method is to construct a numerical scheme that preserves the qualitative features of the continuous system—such as positivity of the solutions and conservation of the total population—regardless of the choice of the time step.

To discretize the time domain, we use a uniform grid with step size $h > 0$, and denote the approximate solutions at time $t_n = nh$ by $S^n \approx S(t_n)$ and $I^n \approx I(t_n)$. Following the principles introduced by Mickens [1], we employ nonstandard denominator functions and appropriate nonlocal approximations to maintain dynamic consistency.

The proposed NSFD scheme for the SIS model is given by:

$$\begin{aligned} \frac{S^{n+1} - S^n}{\phi_1} &= \Lambda - \beta I^n S^{n+1} - \mu S^n + \xi I^n + c_{51}(I^{n+1} - I^n), \\ \frac{I^{n+1} - I^n}{\phi_7} &= \beta I^n S^{n+1} - (\mu + \xi) I^n, \end{aligned} \tag{5}$$

where the denominator functions are defined as

$$\phi_1 = \frac{1 - e^{-h\mu}}{\mu}, \qquad \phi_7 = \frac{1 - e^{-h(\mu+\xi)}}{\mu + \xi}, \tag{6}$$

and the correction coefficient is given by

$$c_{51} = \phi_7^{-1} - \phi_1^{-1}. \tag{7}$$

This scheme was developed based on the exponential integration approach introduced in our previous study [11]. Prior research has successfully applied this approach to SIR, SEIR, and SVIR models, demonstrating that it constitutes a general methodology with broad applicability to more detailed epidemic models [12,13]. Therefore, by extending this methodology to the SIS model, we have established a robust scheme that provides a solid foundation for rigorous theoretical analysis and comprehensive numerical experiments.

4 Theoretical Analysis of the Scheme

In this section, we examine the theoretical properties of the proposed nonstandard finite difference scheme. For a numerical method to be dynamically consistent with the original epidemiological model, it must accurately reflect the fundamental behavior of the continuous system. This includes maintaining the integrity of population dynamics, ensuring that solutions remain biologically meaningful, and providing stable long-term behavior.

The proposed scheme is designed to preserve the total population over time in accordance with the governing demographic processes of birth and death. Additionally, it ensures that all solution variables—such as the number of susceptible and infectious individuals—remain non-negative throughout the simulation, thereby upholding the biological realism of the model. Moreover, the scheme exhibits stable behavior near equilibrium points, meaning that small numerical perturbations do not lead to divergence or spurious oscillations.

Through rigorous analysis in the following subsections, we demonstrate that the scheme satisfies these essential properties. This guarantees that it provides a reliable and qualitatively accurate representation of the original system, independent of the choice of time step.

4.1 Conservation of Total Population

In the SIS model with recruitment and natural death, the total population is not constant over time but follows a predictable dynamic. It is therefore important that the numerical scheme accurately captures this evolution.

Let $P(t) = S(t) + I(t)$ denote the total population at the t-th time step. Summing the two equations of the continuous system (1), we obtain the following differential equation for the total population:

$$\frac{dP}{dt} = \Lambda - \mu P, \tag{8}$$

whose exact solution is given by:

$$P(t) = \frac{\Lambda}{\mu} + \left(P^0 - \frac{\Lambda}{\mu}\right) e^{-\mu t}.$$

Applying the same summation to the discrete system (5), we get:

$$\frac{P^{n+1} - P^n}{\phi_1} = \Lambda - \mu P^n. \tag{9}$$

Theorem 1. *Equation* (9) *is the exact finite difference scheme for the continuous equation* (8).

Proof. We prove by induction that $P^n = P(t^n)$, where $t^n = nh$.

Base case ($n = 0$): The initial condition gives $P^0 = P(t^0)$.

Assume the identity holds for $n = k$, i.e.,

$$P^k = \frac{\Lambda}{\mu} + \left(P^0 - \frac{\Lambda}{\mu}\right) e^{-\mu t^k}.$$

From equation (9), we write:

$$P^{k+1} = (1 - \mu\phi_1)P^k + \Lambda\phi_1.$$

Using the identity $1 - \mu\phi_1 = e^{-h\mu}$, we get:

$$P^{k+1} = e^{-h\mu} P^k + \frac{\Lambda}{\mu}(1 - e^{-h\mu}).$$

Substituting the inductive assumption for P^k, we obtain:

$$P^{k+1} = \frac{\Lambda}{\mu} + \left(P^0 - \frac{\Lambda}{\mu}\right) e^{-\mu(t^k+h)} = P(t^{k+1}),$$

which completes the proof.

This result confirms that the discrete scheme exactly replicates the evolution of the total population as described by the continuous model, ensuring accurate population tracking at every time step.

4.2 Positivity

An essential requirement for any realistic compartmental model of infectious disease transmission is that its solutions remain non-negative over time. In the SIS model, this means that both the susceptible and infected populations must remain greater than or equal to zero for all time steps.

Definition 1 ([14]). *Let $\mathcal{G} = (\mathcal{N}, \mathcal{E})$ be a multidigraph with labeling $\pi\colon \mathcal{E} \to R$. Let $\eta\colon \mathcal{E}^- \to \mathcal{P}(\mathcal{E}^+)$ be a map. The pair $(\mathcal{G}, \eta)$ is an* ***edge partition*** *if:*

(i) $\mathcal{E} = \mathcal{E}^+ \sqcup \mathcal{E}^-$.
(ii) All cycles in $\mathcal{G}$ contain at most one edge in $\mathcal{E}^-$.
(iii) The map η is such that for every $e \in \mathcal{E}^-$:
  (a) if $e' \in \eta(e)$, then $s(e) = s(e')$,
  (b) if $e' \in \eta(e)$, then every cycle containing e' contains $t(e)$,
  (c) if $e \neq e'$, then $\eta(e) \cap \eta(e') = \emptyset$.

We say $\mathcal{G}$ is a ***P-graph*** *if there is an edge partition $(\mathcal{G}, \eta)$ such that for every $e \in \mathcal{E}^-$,*

(iv) $\pi(e) + \sum_{e' \in \eta(e)} \pi(e') \in R_{\geq 0}$.

According to Theorem 3 in [14], if a discrete system admits a P-graph structure as defined above, then the solution of the corresponding linear system remains non-negative for all time steps. In our case, we will show that the scheme (5) satisfies the P-graph condition. Therefore, by this result, it follows that the numerical solution of the scheme is guaranteed to remain non-negative.

Theorem 2. *If the initial values satisfy $S^0, I^0 \geq 0$, then the solution of the scheme (5) remains non-negative for all $n \geq 0$.*

Proof. To prove non-negativity, we rewrite the scheme (5) in matrix form:

$$A^n x^{n+1} + b^n = 0, \quad n = 0, 1, 2, \ldots,$$

where

$$x^{n+1} = \begin{bmatrix} S^{n+1} \\ I^{n+1} \end{bmatrix}, \quad A^n = \begin{bmatrix} -\phi_1^{-1} - \beta I^n & \phi_7^{-1} - \phi_1^{-1} \\ \beta I^n & -\phi_7^{-1} \end{bmatrix},$$

and

$$b^n = \begin{bmatrix} \Lambda + (\phi_1^{-1} - \mu)S^n + (\phi_1^{-1} - \phi_7^{-1} + \xi)I^n \\ (\phi_7^{-1} - \mu - \xi)I^n \end{bmatrix}.$$

Let us now construct the associated Laplacian matrix L:

$$L = \begin{bmatrix} L_{11} \; L_{12} \; L_{13} \\ L_{21} \; L_{22} \; L_{23} \\ L_{31} \; L_{32} \; L_{33} \end{bmatrix},$$

where

$$\begin{aligned} &L_{11} = -\phi_1^{-1} - \beta I^n, && L_{12} = \phi_7^{-1} - \phi_1^{-1}, && L_{13} = \Lambda + (\phi_1^{-1} - \mu)S^n + (\phi_1^{-1} - \phi_7^{-1} + \xi)I^n, \\ &L_{21} = \beta I^n, && L_{22} = -\phi_7^{-1}, && L_{23} = (\phi_7^{-1} - \mu - \xi)I^n, \\ &L_{31} = \phi_1^{-1}, && L_{32} = \phi_1^{-1}, && L_{33} = -\Lambda - (\phi_1^{-1} - \mu)(S^n + I^n). \end{aligned}$$

We construct the corresponding graph shown in Fig. 2.

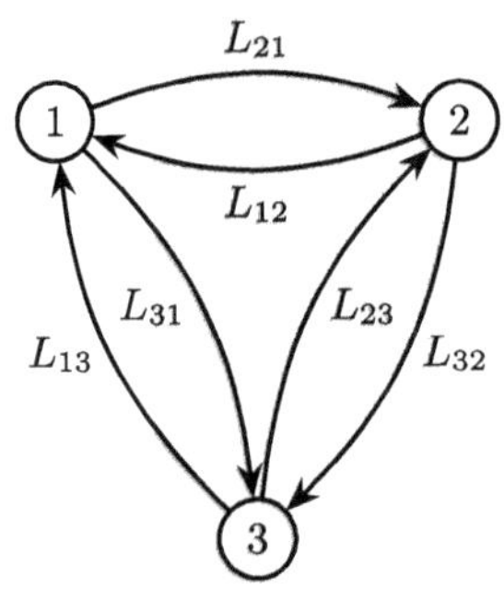

Fig. 2. Graph representation of the SIS model for positivity analysis.

We now analyze the sign of each edge label. Since $\phi^{-1}(h, \lambda)$ is monotonically increasing in $\lambda > 0$, the terms $L_{12}, L_{21}, L_{23}, L_{31}, L_{32}$ are clearly positive.

The term L_{13} requires further attention. The first two components Λ and $(\phi_1^{-1} - \mu)S^n$ are evidently non-negative for $S^n \geq 0$. The last component is $(\phi_1^{-1} - \phi_7^{-1} + \xi)I^n$. To prove this term is non-negative, we invoke the Mean Value Theorem for exponential functions, which implies:

$$0 < \phi_7^{-1} - \phi_1^{-1} < \xi,$$

hence:

$$\phi_1^{-1} - \phi_7^{-1} + \xi > 0.$$

Therefore, L_{13} is also positive.

Since all edge weights are non-negative and $\mathcal{E}^- = \emptyset$, the graph satisfies all conditions (i)–(iv) in Definition 1. Hence, the system matrix corresponds to a P-graph. According to Theorem 3 in [14], this guarantees that the numerical solution of the system remains non-negative at each time step.

4.3 Elementary Stability

In this section, we examine the stability properties of the proposed numerical scheme and provide sufficient conditions under which the equilibrium points of the continuous model (1) are preserved in the discrete setting. Finite difference schemes that maintain the same stability characteristics as the original differential equations are referred to as *elementary stable methods* [15].

Definition 2 ([15]). *The finite difference method is called elementary stable, if for any value of the step size h, its only equilibrium points $\overline{E}$ are those of the original differential system, the linear stability properties of each $\overline{E}$ being the same for both the differential system and the discrete method.*

Lemma 1 ([15]). *For the quadratic equation $\lambda^2 - \Omega\lambda + \omega = 0$, by using the well-known Jury condition, both roots satisfy $|\lambda_i| < 1, i = 1, 2$ if the following conditions are satisfied:*

$$1 + \Omega + \omega > 0, \tag{10}$$

$$1 - \Omega + \omega > 0, \tag{11}$$

$$\omega < 1. \tag{12}$$

Theorem 3. *The proposed scheme* (5) *is elementary stable.*

Proof. The Jacobian matrix J of the system (5) is given by:

$$J(S^n, I^n) = \begin{bmatrix} \dfrac{\partial S^{n+1}}{\partial S^n} & \dfrac{\partial S^{n+1}}{\partial I^n} \\ \dfrac{\partial I^{n+1}}{\partial S^n} & \dfrac{\partial I^{n+1}}{\partial I^n} \end{bmatrix} \tag{13}$$

where

$$\begin{aligned}
\frac{\partial S^{n+1}}{\partial S^n} &= \frac{1 - \mu\phi_1}{1 + \beta\phi_7 I^n}, \\
\frac{\partial S^{n+1}}{\partial I^n} &= \frac{(\mu + \xi - \beta(\Lambda\phi_1 + S^n))\phi_7 - (1 - \beta\phi_7 S^n)\mu\phi_1}{(1 + \beta\phi_7 I^n)^2}, \\
\frac{\partial I^{n+1}}{\partial S^n} &= \beta\phi_7 \frac{\partial S^{n+1}}{\partial S^n} I^n, \\
\frac{\partial I^{n+1}}{\partial I^n} &= 1 - (\mu + \xi)\phi_7 + \beta\phi_7 S^{n+1} + \beta\phi_7 \frac{\partial S^{n+1}}{\partial I^n} I^n.
\end{aligned}$$

Substituting the disease-free equilibrium $E_0^* = (\Lambda/\mu, 0)$ into (13), we get

$$J(E_0^*) = \begin{bmatrix} 1-\mu\phi_1 & (\mu+\xi)\phi_7 - \mu\phi_1 - \dfrac{\Lambda}{\mu}\beta\phi_7 \\ 0 & 1-(\mu+\xi)\phi_7 + \dfrac{\Lambda}{\mu}\beta\phi_7 \end{bmatrix}. \tag{14}$$

The eigenvalues of (14) are:

$$\begin{aligned} \lambda_1 &= 1-\mu\phi_1, \\ \lambda_2 &= 1-(\mu+\xi)\phi_7(1-\mathcal{R}_0). \end{aligned}$$

If $\mathcal{R}_0 \le 1$, then $0 < \lambda_1, \lambda_2 < 1$, implying global stability of E_0^*. If $\mathcal{R}_0 > 1$, then $\lambda_2 > 1$, and E_0^* is unstable.

Now, suppose $\mathcal{R}_0 > 1$. The system admits an endemic equilibrium E_1^*. Evaluating the Jacobian (13) at E_1^* yields:

$$J(E_1^*) = \begin{bmatrix} \dfrac{1-\mu\phi_1}{1+(\mu+\xi)\phi_7(\mathcal{R}_0-1)} & -\dfrac{\mu\phi_1}{1+(\mu+\xi)\phi_7(\mathcal{R}_0-1)} \\ \dfrac{(1-\mu\phi_1)(\mu+\xi)\phi_7(\mathcal{R}_0-1)}{1+(\mu+\xi)\phi_7(\mathcal{R}_0-1)} & \dfrac{1+(1-\mu\phi_1)(\mu+\xi)\phi_7(\mathcal{R}_0-1)}{1+(\mu+\xi)\phi_7(\mathcal{R}_0-1)} \end{bmatrix}. \tag{15}$$

Let

$$a = \frac{1-\mu\phi_1}{1+(\mu+\xi)\phi_7(\mathcal{R}_0-1)},$$

and rewrite (15) as:

$$J(E_1^*) = \begin{bmatrix} a & -a\dfrac{\mu\phi_1}{1-\mu\phi_1} \\ 1-a-\mu\phi_1 & 1-\mu\phi_1 + a\dfrac{\mu\phi_1}{1-\mu\phi_1} \end{bmatrix}. \tag{16}$$

When $\mathcal{R}_0 > 1$, we clearly have $0 < a < 1$. The eigenvalues of (16) are the roots of the equation $\lambda^2 - \Omega\lambda + \omega = 0$, with

$$\begin{aligned} \Omega &= \mathrm{Tr}(J(E_1^*)) = a + 1 - \mu\phi_1 + a\frac{\mu\phi_1}{1-\mu\phi_1}, \\ \omega &= \det(J(E_1^*)) = a > 0. \end{aligned}$$

Since $0 < a < 1$, and all conditions (10)–(12) are satisfied, it follows from Lemma 1 that both eigenvalues lie within the unit disk. Thus, the scheme (5) maintains the same stability characteristics around both equilibrium points as the original system, and is therefore elementary stable. □

5 Numerical Results

In this section, we present several numerical experiments to evaluate the performance, accuracy, and robustness of the proposed NSFD scheme for the SIS model. The proposed scheme is compared with the classical Euler (standard) method, the second-order Runge–Kutta (RK2) method, and the Mickens-type NSFD scheme proposed in [4]. All methods are tested under identical model parameters and initial conditions.

It should be noted that the proposed NSFD scheme differs structurally from the classical and Mickens-type formulations. In the present approach, the discrete derivative is constructed using a nonstandard denominator function $\phi(h)$ that may differ between the equations of the system, allowing better adaptation to the local dynamics of each compartment. Moreover, an additional correction term c is introduced to ensure the consistency between the discrete and continuous formulations, enhancing both the stability and population conservation properties of the scheme. These features distinguish the proposed method from existing NSFD approaches and serve as the main sources of its improved accuracy and robustness. Since the SIS model does not admit a closed-form analytical solution, a reference solution is generated using the standard Euler method with a very fine time step, $h_{\text{exact}} = h/1000$, where h denotes the time step used in each simulation.

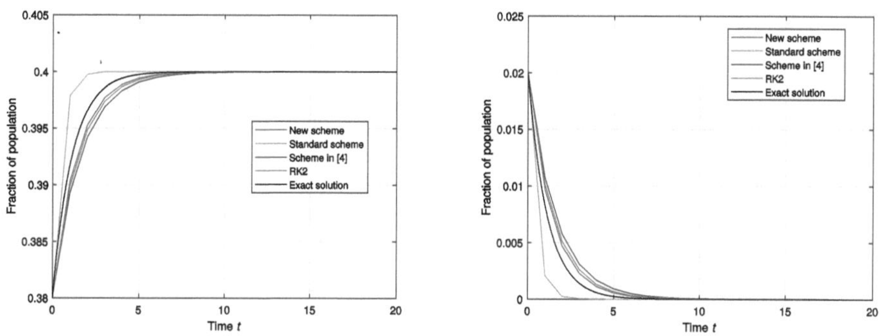

(a) Susceptible population

(b) Infected population

Fig. 3. Comparison of numerical solutions for the SIS model with step size $h = 1$ obtained using the standard Euler method, the Runge–Kutta method, the NSFD scheme in [4], and the proposed NSFD scheme. The model parameters are $\beta = 0.8$, $\xi = 0.7$, $\mu = 0.5$, and $\Lambda = 0.2$, with initial conditions $S(0) = 0.95$ and $I(0) = 0.05$.

Figure 3b illustrates the time evolution of the susceptible and infected populations for $h = 1$, which represents a relatively large time step. It is observed that the proposed NSFD scheme maintains the realistic dynamic behavior of the SIS model and produces solutions that remain close to the reference solution. In contrast, the standard Euler and the Mickens-type NSFD schemes exhibit noticeable deviations or oscillations, indicating reduced stability for larger time steps. This demonstrates that the proposed method is capable of preserving the qualitative features of the continuous model even for coarse discretizations.

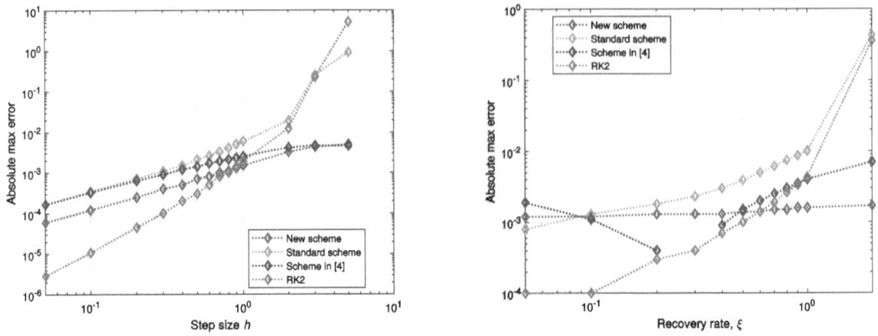

(a) Absolute maximum error vs step size

(b) Absolute maximum error vs recovery rate

Fig. 4. Comparison of the absolute maximum error (in log–log scale) of the numerical solutions for the susceptible population in the SIS model, obtained using the standard Euler method, the Runge–Kutta method, the NSFD scheme in [4], and the proposed NSFD scheme.

To further evaluate the accuracy and stability, the maximum absolute errors of all methods were computed for various time step sizes and parameter values, as shown in Fig. 4.

Figure 4(a) presents the variation of the absolute error with respect to the step size h. The RK2 method achieves the smallest error for very small step sizes, but its accuracy rapidly deteriorates as h increases, showing high sensitivity to the step size. In contrast, the proposed NSFD scheme exhibits the smallest dependency on h, maintaining stable accuracy across a wide range of step sizes. This behavior clearly highlights the robustness of the proposed scheme with respect to the discretization parameter.

Figure 4(b) shows the effect of varying the recovery rate ξ on the absolute error. The results indicate that the proposed NSFD method is the least sensitive to changes in ξ, providing stable and consistent accuracy compared to the other methods. This demonstrates that the method remains reliable under parameter uncertainty, an important feature for epidemiological modeling where parameters may not be precisely known.

Additional simulations were performed for different parameter sets (β, Λ, μ) and various initial conditions (S_0, I_0). In all cases, the proposed NSFD scheme successfully preserved the population invariant $S + I = \Lambda/\mu$, while ensuring the positivity and boundedness of the numerical solutions. These results confirm that the proposed scheme is not only accurate and stable but also robust and dynamically consistent, making it a reliable numerical tool for long-term simulations of SIS-type epidemic models.

6 Conclusion

In this study, we have developed a novel NSFD scheme for the SIS epidemic model and evaluated its theoretical and numerical performance. On the theoretical side, the proposed scheme has been rigorously proven to satisfy the fundamental qualitative properties of epidemic dynamics, including the *conservation of total population*, *positivity of the solutions*, and *elementary stability* of equilibrium points. These properties ensure that the discrete model preserves the essential dynamical behavior of the continuous system and remains stable over long time intervals.

The numerical experiments demonstrate that the proposed NSFD method outperforms the standard Euler method, the second-order Runge–Kutta method, and the Mickens-type NSFD scheme from the literature in terms of stability, robustness, and overall accuracy. The results show that our scheme provides stable and realistic dynamics even for relatively large time steps, while maintaining weak sensitivity to parameter variations. In contrast, although the RK2 method achieves higher local accuracy for small step sizes, it fails to guarantee population conservation and the positivity of solutions, highlighting that high-order accuracy does not necessarily imply dynamically consistent behavior.

Overall, the proposed NSFD scheme preserves the qualitative features of the SIS model for all step sizes and parameter choices, producing reliable and stable numerical solutions that accurately reproduce the true dynamics. These findings suggest that the present approach offers a promising and robust framework for the numerical simulation of not only SIS-type models but also more complex epidemic systems with similar structural characteristics.

Competing Interests. The authors declare that they have no competing interests.

References

1. Mickens, R.: Nonstandard Finite Difference Models of Differential Equations. World Scientific, Singapore (1994)
2. Mickens, R.: Nonstandard finite difference schemes for differential equations. J. Differ. Equ. Appl. **8**(9), 823–847 (2002)
3. Hoang, M., Egbelowo, O.: Nonstandard finite difference schemes for solving an SIS epidemic model with standard incidence. Rend. Circ. Mat. Palermo Ser. **II**(69), 753–769 (2020)
4. Roeger, L.-I.W.: Dynamically consistent discrete-time SI and SIS epidemic models. Discrete Contin. Dyn. Syst. **33**(2), 653–662 (2013)
5. Agus, S.: A nonstandard finite difference scheme for sis epidemic model with delay: stability and bifurcation analysis. Appl. Math. **3**, 528–534 (2012)
6. Gray, A., Greenhalgh, D., Hu, L., Mao, X., Pan, J.: A stochastic differential equation SIS epidemic model. SIAM J. Appl. Math. **71**(3), 876–902 (2011)
7. Cai, S., Cai, Y., Mao, X.: A stochastic differential equation SIS epidemic model with two correlated Brownian motions. Nonlinear Dyn. **97**(4), 2175–2187 (2019). https://doi.org/10.1007/s11071-019-05114-2
8. Rifhat, R., Ge, Q., Teng, Z.: The dynamical behaviors in a stochastic SIS epidemic model with nonlinear incidence. Comput. Math. Meth. Med. (2016)
9. Saito, K., Kimura, M., Ohara, K., Motoda, H.: Efficient discovery of influential nodes for SIS models in social networks. Knowl. Inf. Syst. **30**, 613–635 (2012)

10. Van Mieghem, P.: The N-intertwined SIS epidemic network model. Comput. **93**, 147–169 (2011)
11. Enkh-Amar, S., Batgerel, B.: Extending nonstandard finite difference scheme for SIR epidemic model. Springer Proc. Math. Stat. **434**, 187–200 (2023)
12. Enkh-Amar, S., Batgerel, B.: Extending nonstandard finite difference scheme for the SEIR epidemic model. J. Inst. Math. Digit. Technol. **4**, 16–24 (2022)
13. Tumurkhuyag, B., Batgerel, B.: Nonstandard finite difference scheme for the epidemic model with vaccination. J. Math. Sci. **279**, 841–849 (2024)
14. Saez, M., Feliu, E., Wiuf, C.: Graphical criteria for positive solutions to linear systems. Linear Algebra Appl. **552**, 166–193 (2018)
15. Khalsaraei, M.M., Shokri, A., Noeiaghdam, S., Molayi, M.: Nonstandard finite difference schemes for an SIR epidemic model. Mathematics **9**, 3082 (2021)

Author Index

A. Chinchuluun et al. (Eds.): AAIM 2025, LNCS 16379, p. 189, 2026.
https://doi.org/10.1007/978-981-95-5657-1

MIX
Papier aus verantwortungsvollen Quellen
Paper from responsible sources
FSC® C105338

If you have any concerns about our products,
you can contact us on
ProductSafety@springernature.com

In case Publisher is established outside the EU,
the EU authorized representative is:
Springer Nature Customer Service Center GmbH
Europaplatz 3, 69115 Heidelberg, Germany

Printed by Libri Plureos GmbH
in Hamburg, Germany